FROM MOVABLE TYPE PRINTING
TO THE WORLD WIDE WEB

From Movable Type Printing to the World Wide Web

Essays on Civilizations, Cultures and Education

H K Chang

City University of Hong Kong Press

First published 2007
Second printing with minor amendments 2007
Printed in Hong Kong

ISBN: 978-962-937-143-2

Published by
City University of Hong Kong Press
Tat Chee Avenue, Kowloon, Hong Kong

Website: www.cityu.edu.hk/upress
E-mail: upress@cityu.edu.hk

Dedicated to
Dr. En-shu Chang
My Late Father

Table of Contents

Preface ix

The Author xi

Photo Section xiii

1. Thoughts by the Bank of the River Nile 1

2. Thoughts Along the Seine 11

3. Thoughts by the Ganges 45

4. Thoughts Along the Bosporus Strait 63

5. From *The Adventures of Hajji Baba*
   to *My Name Is Red*: A Personal Journey 83

6. "The Accidental Tourist" in Egypt 105

7. From Zhang Qian's Western Expedition
   to the "Factory of the World" 111

8. The World at the Time of Zheng He 121

9. From Movable Type Printing
   to the World Wide Web 145

10. From Geometry to Tap Dance 159

11. A Dose of Idealism 163

12. SAR's Version of Cultural Fusion 167

13. On Models of Cultivating Creative Talent 171

14. The Brave New World of Learning.................................185

15. A Leading University in the
    Asia-Pacific Region 193

16. The World Wide Web...199

17. Extensive Study, Accurate Inquiry,
    Careful Reflection and Clear Discrimination..........203

18. After Extreme Bad Luck Comes Good Luck...........207

19. Embrace Changes and Exert Yourselves....................213

20. Picture Out of Your Life...219

21. The World is Flat 225

Preface

I was born in Shenyang under the puppet Manchurian regime; had my first lesson in reading in Jinan after China's victory in the Sino-Japanese war; grew up in Taiwan under Chiang Kai-shek's authoritarian rule; formed my own family in one of the bases for the civil-rights movement in the Bay Area of California; and established my career in the State of New York during the Vietnam War.

In the summer of 1976, I moved my family to the Olympic Games host city of Montreal in Canada. The Parti Québécois took over the provincial government and started to call for independence shortly after. We moved again in 1984 to another Olympic Games host city, this time Los Angeles, California, for me to take up a position at the University of Southern California. And in 1990, at the age of fifty when, according to Confucius, I was supposed to have come to know the existence of the heavenly principles, I followed my heart to Hong Kong and embarked on my new journey.

At the age of five, father took me out to the streets to welcome the Chinese troops when they took over Jinan from the Japanese. This was my first encounter with the concepts of nation and state. At the age of ten, father bought me several books written for youngsters. They included stories of national heroes and biographies of famous individuals in the world and the wonders of the world. These books really broadened my horizons. Father was a surgeon by training, but he was quite knowledgeable about geography, history and ethnology. His insights inspired me.

When I went to the U.S. for further studies at the age of

twenty-three, I had the opportunity to tour Asia, Africa and Europe because of my parents' work in the ancient country of Ethiopia in eastern Africa. I saw the Nile, the river that nurtured the ancient Egyptian civilization, for the first time. There I had an initial understanding of the rise and fall of great civilizations in the world. Due to this experience in my youth, I paid special attention to the humanities and social developments throughout the following three decades while I studied and taught biomedical engineering.

During the roughly fifteen years working in Hong Kong, I wrote in both English and Chinese about issues such as the positioning of Hong Kong, higher education, the integration of science and humanities, the development of Chinese culture and the advancement of world civilization. Twenty-one articles have been selected for this book as a record of the intellectual and sentimental journey of mine. The book is titled *From Movable Type Printing to the World Wide Web* because an article of the same title in the book is representative of the twenty-one articles here; it was based on the lecture I gave in 1998 at the inauguration of the Cultural Lecture Series of the Chinese Civilization Center of City University of Hong Kong.

After retirement my parents lived in California. When I informed them in the autumn of 1989 of my decision to go to work in Hong Kong, my father pondered a while and offered his unselfish encouragement for me to leave California and go to Hong Kong where my bicultural background and bilingual abilities would be a definite asset. His encouraging words have given me strength ever since then. Now, with reverent piety, I respectfully dedicate this book to my late father.

H K Chang
March 2007

The Author

Born in Shenyang, China, H K Chang received his Bachelor's Degree in Civil Engineering from National Taiwan University in 1962, his Master's Degree in Structural Engineering from Stanford University in 1964, and his PhD in Biomedical Engineering from Northwestern University in 1969. He has served as a faculty member at the State University of New York at Buffalo and McGill University in Montreal; he was Professor and Chairman of Department of Biomedical Engineering at the University of Southern California in Los Angeles from 1985 to 1990. In 1990, he became the founding Dean of the School of Engineering at the Hong Kong University of Science and Technology and then in 1994, the Dean of the School of Engineering of the University of Pittsburgh. He served as the President and University Professor of City University of Hong Kong from 1996 to 2007.

A world renowned biomedical engineering expert, Professor Chang served in 1988–89 as President of the Biomedical Engineering Society (U.S.). He is a Foreign Member of The Royal Academy of Engineering (U.K.). He has published more than one hundred scientific articles, is the editor of two research monographs and holds a Canadian patent.

He actively promotes local arts activities and cultural exchanges between Hong Kong, the Chinese mainland, and other countries. In 2000, the President of France awarded him France's highest order of honor by appointing him a Chévalier dans l'Ordre National de la Légion d'Honneur in recognition of his academic accomplishments and contributions to cultural exchange between China and France.

Between 2000 and 2003, he was Chairman of the Cultural and Heritage Commission and, between 2000 and 2004, a member of the Council of Advisors on Innovation & Technology, of the Hong Kong SAR. In 2002, he was awarded a Gold Bauhinia Star by the Government of the Hong Kong Special Administrative Region in recognition of his contributions to education, culture, science and technology.

He was appointed an Honorary Professor of Peking University in 2006.

Photo Section

Family photo taken during the writer's sabbatical leave at Université Paris XII in 1982.

At a "City Cultural Salon", a monthly gathering at the writer's residence, 1999.

The Great Egyptian Sphinx, sited on the Giza Plateau of the River Nile, 2005.

◀ The City Palace Museum in Udaipur in the Indian State of Rajasthan, 2006.

▼ With children in a slum in Lucknow, India, 2006.

Along the banks of the Seine in Paris. (Photo by H K Chang, 2006)

A view of Istanbul from the north side of the Golden Horn. (Photo by H K Chang, 2006)

Orhan Pamuk, giving a lecture at the City University of Hong Kong, 2004.

Giving a lecture on "Zhang Qian's Western Expedition to the 'Factory of the World'", Peking University, 2006.

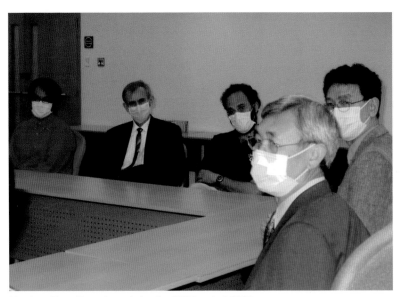

Meeting with staff members during the SARS period, 2003.

At the 20th anniversary celebration, 2004.

Awarded a Gold Bauhinia Star by the government of the Hong Kong SAR, 2002

With Zhang Yimou, Henry T. Yang, Rita Fan Hsu Lai-tai and Ronald Joseph Arculli,
the honorary award holders, at the 2005 Congregation.

Orientation with new CityU students, 2006.

FROM MOVABLE TYPE PRINTING
TO THE WORLD WIDE WEB

1

Thoughts by the Bank
of the River Nile

A Journey of Eight Thousand Li
꒦꒷꒦꒷꒦꒷꒦꒷꒦꒷꒦꒷꒦꒷꒦꒷꒦꒷꒦꒷꒦꒷꒦꒷꒦꒷꒦꒷꒦꒷꒦꒷꒦꒷

Hotel El-Gezirah in Cairo, many many miles from Hong
Kong, is on the southern tip of an island in the Nile. From
here, you can see stretched before you the Nile and the cities
on its two banks.

Sitting in the hotel coffee shop, I watch the water of the
Nile that has nourished Egyptian culture for six thousand
years rush towards me and then continue to flow by me. My
mind turns on its own to thoughts of the past and the present,
bringing back memories of my days at Radio Television Hong
Kong as the host of the show "Chang Hsin-kang's Random
Thoughts." Time and place seem to be all a jumble in my
mind.

This is my second time in Cairo. In 1963, I was getting
ready to leave Taiwan to study in the U.S. My parents were
working for the World Health Organization (WHO) in
Ethiopia in East Africa at the time, and my younger brother
and sisters had gone to the U.S. earlier to study. As I was the
only person in the family left in Taiwan, it would have been
very difficult for me to obtain a student visa to the U.S. That is
why I had flown over Asia, Africa, and Europe, and from
there, over the Atlantic, before traveling across the North
America continent to get to the west coast of the U.S. to enroll
at Stanford University.

My journey took me first to Hong Kong, Bangkok, Bombay, Beirut, and Cairo in North Africa. From there, I passed through Addis Ababa, the capital of Ethiopia, to the ancient capital of Gonder in the northwest region of Ethiopia, where my parents worked. Ethiopia is an ancient country with three thousand years of history. It looks over the Red Sea to the Arabian Peninsula, and is connected to Egypt by the Nile. The kings of this country had always claimed that they were descendants of Queen Sheba and King Solomon. Legend has it that the beautiful Queen Sheba went to visit King Solomon in Jerusalem, and later bore him a son, who became the founder of Ethiopia. Gonder is not far from the source of the Blue Nile, one of two major tributaries of the Nile. I once took a car to the hilly country nearby and looked down at the source of this world-famous river. You could say that I revered or even worshipped the Nile, but I had actually learned only isolated facts and lacked a historical sense to comprehend what I was seeing; it was as if I was able to mumble a few lines of poetry but knew not what they signified. I can only recall that when I looked at the source of the Nile, I murmured lines of Confucius, "It passes on just like this, not ceasing day or night," and the poem "I live by one end of the Long River, and you by the other end. I think of you day by day. I long to see you, but in vain. We drink by the same river . . . "

Two incidents were etched deeply in my memory in the month or so that I stayed in Gonder.

The first incident concerns an Israeli doctor whom I met there. His parents were Russian Jews, but he had been born and had grown up in Harbin until leaving China at the age of eighteen. His work was to take care of the Jews living around Gonder, who were no different from Ethiopians in their skin color, facial features and life-style. This doctor often drove from village to village to treat patients, but he also took the opportunity to track down these Jewish villages one by one. In 1974, a military coup broke out in Ethiopia, overturning the monarchy and putting in its place Soviet-style socialism. Soon after, Israel conducted a carefully planned operation, airlifting more than twenty thousand Ethiopian Jews to Israel. There is no doubt that this doctor had played a fundamental role in the success of the operation.

Now the second incident. There was an Iranian doctor in the Medical College of the WHO where my father taught. His daughter, who was studying in a university in the U.K., was also visiting her parents during the summer vacation. One time, she asked me and several other people to go horseback-riding. Having grown up in Taiwan, I could not say I had touched a horse before. While the other people galloped away, I did not even know what to do with the reins. Fortunately, this young Iranian woman patiently coached me. The next day, my buttocks hurt unbearably, and there were a few places on my inner thighs where the skin had come off. I never heard from her after that brief encounter. After the 1979 Iranian revolution, I often wondered what had happened to her family. She was then a university student, a graceful dancer, and a competent equestrian, but she must well be over sixty years old now.

After I got a visa to go to the U.S., I said goodbye to my parents and continued with my journey from Asmara, which was still part of Ethiopia then, but is now the capital of Eritrea. I passed through Khartoum in Sudan and flew to Athens. From there, I toured and stayed briefly in Zurich, Rome, and Paris before flying to New York, where I got on a Greyhound bus, which was a very popular means of transportation then. In this way, I crossed North America in three days and three nights and arrived in San Francisco in time to report to the university.

My roommate at Stanford was an American graduate student in anthropology. He was very interested in Egypt, and had planned to go there for field work. I learned quite a bit about Egyptian history and culture from him. I had just left Taiwan then, and saw myself as the true inheritor of Chinese culture. Naturally, I could not help from expounding on Chinese culture and history to him. A year later, he decided to take Chinese lessons and got ready to go to Taiwan for his doctoral research.

Forty years quietly went by me, just like the water of the Nile. Last year, Professor David Jordon, an internationally renowned anthropologist and an authority on Chinese folk beliefs, retired from the University of California at San Diego. I had given him his Chinese name, Jiao Dawei, which he had used all these years.

Forty years ago, I may have inadvertently deprived the world of an anthropologist with a specialization in Egyptian studies, but I had also indirectly brought to the world an excellent scholar who is erudite in Chinese culture and full of good wishes for the Chinese people. In forty years, I, too, have evolved from a fortunate foreign student to become, first, the dean of an engineering school in an American university, and, then, as the kind compliments of many Hong Kong people would have it, a "cultured university president."

The Pharaoh who lies in the pyramids just a few kilometers away may offer incontrovertible proof that individual success and wealth in this world are "like fog and lightning." Only objective accomplishments and knowledge worthy of being passed on to later generations can stand the test of history.

Yue Fei of the Song Dynasty once wrote, "At thirty years of age, my accomplishments are like dust and dirt. My sojourn has taken me over eight thousand *li*." Eight thousand *li* cannot even begin to measure the distance I covered in my youth when I traveled to the U.S. for my studies; later, when I decided to come to work in Hong Kong during middle age, I had few regrets at disrupting my career of close to thirty years in North America. Yue Fei would probably nod his head in approval from his grave at this admirer of his who has come a thousand years after him.

Flying to the Homes of the Humble Folk

When they were alive, the Pharaohs of ancient Egypt claimed to be the incarnation of gods. To ensure their eternal comfort in the after-life, they spared no expense and mobilized thousands of slaves to build their graves. These graves were so intricately designed, and the artisans who built them were skillful, that those who set eyes on them cannot but be awed by their grandeur. In the third century B.C.E., Alexander the Great conquered Egypt. His general, Ptolemy, and Ptolemy's descendants after him, ruled Egypt as outsiders. In order to win the acceptance of the local people, the Greeks adopted Egyptian clothing and religion, and constructed many grand temples. The last ruler of the Ptolemy dynasty was Cleopatra, an empress not known for her beauty but for her political acuity. She had such a fondness for silk robes that she was willing to purchase silk from far away China with an equal weight of gold. She was married at different times to the Roman Caesar and Anthony and gave birth to four children. This drama of love and politics became known to many people of the twentieth century, thanks to the Hollywood production and the outstanding performance of Elizabeth Taylor.

In retrospect, the unfair distribution of resources is a

phenomenon found in all societies in human history. The rulers and the upper crust of society have the lion's share of the resources at their command, while the majority of people at the bottom cannot even be assured of having enough food and clothing, and some are even reduced to lives of slavery. Cleopatra had her silk robes, but many of her people did not have enough clothes to cover their bodies.

In fact, what has gradually closed the gap caused by extreme differences in lifestyles among people is not charity on the part of people living at the top, for charity can only solve the immediate crisis but not reduce long-term inequality. Neither is it the result of revolutions that aim to redistribute wealth, because after most revolutions, rulers might change, but not the distribution model itself.

What truly has abolished the slavery system and raised the living standards of ordinary people is the development of industry and the advancement of technology. In the twentieth century, beginning with the U.S., the capability of large-scale industrial and agricultural production in conjunction with market expansion has made it possible for people to enjoy a style of life that only royalty could afford in the past.

Only several decades ago, I was fortunate enough to travel around the world by air. In those days, most people, including many with substantial means, did not have the opportunity to fly. Many families did not even have a telephone at home. Today, one can find tour groups from Taiwan, Hong Kong and the Mainland of China in the Nile Valley and by the side of the pyramids, which goes to show the effects of technological advancement and economic development on our lives.

While we are on the subject of tourism, the Egyptians are truly remarkable. They have the wherewithal to train a sizeable corps of tour guides who are well-steeped in Egyptian history and culture and proficient in Chinese. This is attributable to their higher education system which has produced four Nobel laureates.

The University of Cairo, where I am visiting now, was established in 1908 despite the objections of the British, the rulers of the time. It is the best university in Egypt, with twenty-three schools, one hundred and ninety thousand regular students, seven thousand and eight hundred faculty members, and eight thousand assistant instructors. The administration building of another university in Cairo, Ain Shams (sun in ancient Egyptian) University, is located in the palace where the last emperor of Egypt, King Farouk (who abdicated in 1952), was born. It has one hundred and seventy thousand students, and a superb Chinese Department with a fairly long history. In addition, Cairo is home to the American University in Cairo, established by Americans, which has produced many remarkable individuals in recent years.

Just like in China, where old-style academies such as the Yuelu Academy and Donglin Academy stood before the advent of the modern university, the Al-Azhar Mosque in Cairo has an academy (Madrasa) devoted to the studies of the classics in Islam. In the last thousand years, it has nurtured numerous Sunni scholars. Ibn Khldun, the most important historian in the Islamic world, taught and did research at this institute. He studied the history of various dynasties and cultures, and formulated a law of history, namely, that all civilizations and dynasties undergo a cycle of prosperity and decline. Such a cyclical view of history influenced the twentieth-century English historian Toynbee, who argued that the twenty-first century should belong to Asians in the Pacific region.

Quietly, the Nile flows, while the wheel of history rolls on. Only a small elite could benefit from a university education in the past, but today, the children of ordinary people can attend university. Only an elite minority could have access to culture, taste and manners in the past; should that still be the case today?

In recent years, I have argued for placing equal emphasis on the study of technology and the humanities, have lectured at many universities on the mainland and in Taiwan, and have published my views on the topic. My colleagues and people from all quarters of society have given me strong encouragement and support. During this visit, as I have examined Egyptian higher education, I have also told my Egyptian colleagues of the educational ideal that I have embraced for many years, as well as my experience in putting it into practice.

Liu Yuxi of the Tang dynasty had a famous line, "The swallows nesting in front of the grand halls of the Wangs and Xies have now all flown to the humble homes of the ordinary people." His original meaning was that wealth and power such as those enjoyed by the prominent clans of Wangs and Xies during the Six Dynasties cannot be counted on to endure, but I believe that progress in history means precisely that the humble ordinary people will also come to see the splendor of the rich.

The goal of education includes the promotion of culture, taste and social graces. In order to increase vertical social mobility, culture, taste and social graces should not be monopolized by the elites. The overwhelming majority of students at City University come from humble ordinary families. In the last ten years, I have devoted no small effort to promoting culture in and out of the classroom, with the sole purpose of bringing these students closer to the grand halls of the Wangs and Xies. Ten years of dedication, a shock of white hair – all because of that.

The Hawk and the Peng Bird

Among the ancient Egyptian deities is Horus, who is represented by the head of a hawk, the drawing of which has made its way onto the badge of the modern Egyptian police and army.

For China, the totem animal is the dragon, and it seems we do not pay much heed to the hawk. The closest reference is the *peng* bird in Zhuang Zi's "Free and Easy Wandering." Nobody can tell how big a *peng* bird is or how high and far it can fly. That, however, does not stop us from wishing people "flying ten thousand *li* like a *peng* bird" or using the expression to spur ourselves on to high achievements.

To be sure, life is not a smooth sailing trip. One therefore has to learn to accept success and failure in teaching our students and training ourselves. A Russian adage puts it this way, "A hawk (or an eagle) may fly lower than a chicken, but a chicken can never hope to soar as high as a hawk (or an eagle)." We would all do well to remember this saying.

As I look at the portrait of the hawk head of Horus on this eve of 2006, my thoughts are with you, my friends, many miles away. May all of us soar high, as high as a hawk, and fly as far as a *peng* bird!

(January 2006)

2

Thoughts Along the Seine

I Came, I Saw, I Was Conquered

I was eleven years old when I first developed an interest in France. In a civic lesson in my grade school in Taipei, the teacher asked us to arrange our desks in a circle, and we held a mini-United Nations Security Council meeting that we rehearsed earlier. I was assigned to be the French ambassador and, from then on, developed a fondness for France. When I attended National Taiwan University, I devoted all sixteen credit units of my free electives throughout my four years to studying French.

I went to France for the first time in August of 1963. I had just said goodbye to my parents, who were working for the World Health Organization in Ethiopia, to go to the U.S. to study. I started off from Addis Ababa, went through Athens, Rome and Zurich, and arrived in Paris. Because the bank in Zurich would only allow me to deposit, not cash, the American check I had with me, I had only US$30 on me when I got to Paris. Luckily, I found out from the information desk at the airport about a family-style bed and breakfast (*pension*) on the left bank of the Seine that charged only US$3.50 a day, making it possible for me to roam on my own in Paris for three days with only those thirty dollars.

In three days, with the help of a map and my broken French, I visited all the must-see sites for tourists, feeling quite happy with myself for having learnt a few things. Especially after my tour of the Louvre, I felt like the young Wang Mian in *The Scholars* who, after reading on his own for a few years, seemed "to have got it!"

When I visited the Louvre again, it was fourteen years later. In 1977, at the invitation of International Union of Physiological Sciences, I gave a special address at their conference held in Paris. My memory of the five days of the conference is rather vague, except for the time when I made my report at Paris VII and the big reception hosted by Mayor Jacques Chirac (now President of France) at the newly opened Pompidou Center, but I can still see clearly all the sights and activities that I sneaked out from the conference to enjoy, along with the avante garde theater that I saw at an arts festival in the famous city of Avignon in southern France after the conference was over.

In 1979, I began my collaboration with a laboratory at Paris XII that would last for eleven years. I accepted a number of graduate students and post-doctorates from that institution, and on two occasions, two of my doctoral candidates also went to Paris to conduct experiments using the equipment available there.

The most meaningful period in those eleven years of collaboration came when I was on a twelve-month academic leave from McGill University in Canada in 1981–82, which allowed me to take up a one-year visiting professorship at Paris XII. The four of us rented a two-story house with a courtyard in the suburb of Créteil southeast of Paris not far from Marne, a tributary of the Seine. My daughter was enrolled in a local secondary school as a first-year student, and my son attended the fifth grade in a primary school nearby. The community was made up of businessmen, civil servants, staff members of the university hospital, and immigrants from north Africa who were primarily workers. We seemed to be the only Chinese family around.

During that year, apart from finishing a number of important academic articles, my most rewarding experience was the friendship that I formed with a group of French nationals. I was also able to gain a deeper understanding of the way of life and state of mind of the average Parisian by observing my children's friends and participating in the activities of the parent group.

In France, there is a week of recess after every six weeks of class in high school and grade school, an arrangement that we took advantage of by driving to tour various parts of France. There were a few times that we checked out of an inn in the morning with maps and guidebooks in hand, not knowing what our next stop would be, let alone where we were going to spend the night.

In 1990, my wife and I came to work in Hong Kong. I have had many fewer opportunities to read and write in French, but our fondness for France has in no way diminished. In the last ten years or so, my wife and I have taken many vacations to France, where we have caught up with our friends. On two occasions, our whole family also drove around towns rarely visited by foreign tourists. We sometimes converse in a mixture of Chinese, English and French in restaurants, which makes us a curiosity to local people. On one occasion, this led to an invitation from a retired teacher to join him for red wine and cheese at his home.

In May of 2006, I visited Paris for close to a month in preparation for a program on world civilization. I saw many of my old friends that time. I was most touched by a family dinner, hosted by Professor Laurent and his wife, with nine people attending. Professor Laurent was the chair of the department when I arrived at Paris XII as a visiting professor in 1981. (He was appointed to be the President soon after I arrived.) Then there were the newly retired Professor Atlan and his wife; Dr Isabey, my post-doctorate student, and currently the director of the laboratory, and his wife; and Professor Lancner, an expert in medieval French literature at Paris III. She is the widow of Professor Harf, with whom I had enjoyed the best working relationship and deepest friendship. He visited my laboratory at McGill and the University of Southern California a few times, but he had died three years earlier at the height of his career. Professor Laurent, now in his eighties and retired from the Presidency, invited all of them for a gathering at his home, and all of them came. We had many happy memories of our twenty-five-year friendship, but could not but feel a strain of sorrow that one of our friends was no longer there to share our company.

Many people say that the French, for all their good manners, are in fact quite arrogant, seldom inviting friends from other countries to their homes and being slow to form deep friendships. I can tell from my own experience that this is not the case.

Language barriers and cultural prejudices often create obstacles in human interaction. But, as soon as people become trusted friends, differences of nationality pose no impediments to the depth and truthfulness of their friendship. On that occasion in Paris, I felt profoundly that, like red wine, true friendship ages well.

Fifty-five years have passed since I made my acquaintance with France, and it was forty-three years ago that I visited France for the first time.

Two thousand years ago, Caesar, who had warred in Gaule (France today), made a famous victorious declaration: "Veni, vidi, vici" (I came, I saw, I conquered). If someone were to ask me for my overall impression of France after my many visits from 1963 to 2006, I would say, after Caesar, "Veni, vidi, victus sum." (I came, I saw, I was conquered.)

A Moveable Feast

In 1986, I came across Hemingway's *A Moveable Feast,* in which he records vignettes of his life in Paris from 1921 to 1926. The name and content of this book set it apart from the abundance of books about Paris. It is also a model of modern English prose. With its crisp and economical language, it succeeds in capturing in a vivid way humans and their affairs.

In May of this year, I visited Paris I. The university set aside an office next to that of the First Vice-President for my use. Her office and the Foreign Student Affairs Office (there are eight thousand foreign students on campus) were housed in a luxurious mansion converted from a former private residence, which is located about one and a half kilometers southeast of the main campus. I was staying in a hotel one kilometer northwest of the campus, and walked to and from my office every day.

One day, a thought came to me. All of these places that I walked by every day, I thought – aren't they the same as those Hemingway describes in *A Moveable Feast?* I checked out the book from the library and read it through again, getting a better understanding of the world in which Hemingway lived in those days.

One of Hemingway's essays is entitled "People of the Seine." There would be no Paris without the Seine. Around 300 B.C.E., a fishing tribe named the Parisii (from which Paris gets its name) settled on the island in the Seine. The island was Ile de la Cité, on which the Notre Dame stands. Today, all mileage markings on French national highways are measured from this point.

In 55 B.C.E., Roman troops conquered Gaule (today's France) and came to Paris. They were stationed on the left bank of the Seine, and constructed hot spring bathhouses, amphitheaters and the other structures that have come to represent Roman civilization. In the last century or so, archeologists unearthed many relics from the past. Cluny, which Hemingway mentions a number of times in his book, was the site of a Roman hot spring. The Fifth Republic had just begun when Hemingway died. He and the lost generation (*la generation perdue*) to which he belonged could not have foreseen the prosperity and progress of the Fifth Republic, whose several presidents, incidentally, did a lot to restore these sites. The Musée National du Moyen Age that was completed recently is in Cluny. It preserves a hot spring from the Gallo-Roman period and other relics, and a medieval structure and courtyard on its site has been restored. The elegant design of the museum preserves the spirit of the original structure, and the exhibition inside is rich and superb.

A few minutes' walk from Cluny brings one to one of the oldest universities in Europe – la Sorbonne. The famous Café de Cluny is just nearby, where numerous men of letters have fed their bodies and stimulated their minds throughout French history. Twenty-five years ago, I, too, came by this place for knowledge. Professor Lancner, wife of my deceased friend Professor Harf, completed her National Doctoral dissertation in 1981. I attended the oral presentation of her dissertation. The four-hour event was broadcast on FM radio. Afterwards, we retired to Café de Cluny for a celebrative reception. While most of the guests were engaged in a discussion comparing English and French medieval literature, I helped myself to the delicious food kindly provided by my hosts, in playful observance of Confucius' dictum that "when it comes to food, the master should be the first one served."

The Sorbonne was built in the St. Louis (Louis IX) period of the thirteenth century, a time when the crusaders from Western Europe met the Mongolian troops in today's Middle East, and when Western Europe was beginning to emerge from the benighted Middle Ages. In Catholicism, the enlightenment can be seen in Thomas Aquinas, whose scholasticism incorporates the thought of Plato and Aristotle of ancient Greece. The Sorbonne, therefore, can be said to have taken the lead in liberating France from the dark ages, beginning with the teaching of theology and later adding onto the curriculum such disciplines as literature, law, and medicine. Gradually, it became an academic center of the Renaissance in Europe. Because Latin-speaking monks and students inhabited in the neighborhood, the area is known as *Quartier Latin* (the Latin Quarter).

Today, in addition to the Sorbonne, the Latin Quarter is filled with all kinds of bookstores. Hemingway wrote about Shakespeare & Company, which sold as well as rented books. Hemingway was able to obtain many books from the kind owner, apparently on credit. Today, the bookstore is a disorderly hole in the wall, but every book that is sold from the premises is stamped with a Shakespeare & Company seal that Hemingway made popular. Perhaps that is the way the bookstore wishes to recoup the debt from Hemingway.

In those days, Hemingway and his wife rented a penthouse on rue du Cardinal Lemoine. In winter, they burned oval-shaped pieces of coal for warmth. He walked many a time through a long and narrow commercial street called rue Mouffetard. I, too, had to walk through the same street from the campus of Paris I to my office. Eighty years ago, Hemingway saw squat toilets on the two sides of the street and a horse cart would come to collect the night soil from these toilets every morning. Twenty-five years ago, a French friend of mine recommended that I come here to try out the delicious crepes, especially the Brittany galette, apparently the best in France. On my last visit, I came to this place again, and discovered that the eateries are now mostly run by Greeks and Arabs who serve seafood and shish kebab. For crepes, one would now have to go to Montparnasse, where many cultural celebrities of the nineteen-thirties lived and where Hemingway himself frequented.

A Moveable Feast is indeed movable. Eighty years ago, one would have been hard pressed to find a Chinese restaurant in Paris, let alone a Chinatown. In today's Paris, no matter what district you are in, you will find many Chinese restaurants. Thirty years ago, a large contingent of immigrants of Chinese heritage from Vietnam and Cambodia began to settle near Porte de Choisy in the 13th District and they opened quite a few restaurants specializing in soup noodles. Now, you can find practically anything – Chinese grocery stores, department stores, tourist agencies, and beauty parlors. But the Seine has caught on, and the Porte de Choisy is now known as the Old Chinatown. In the last ten years, immigrants from the mainland of China have moved in droves into Belleville in the 19th District, and a new Chinatown is taking shape. On a May evening, I walked through the area, and noticed Chinese living side by side with Arab immigrants from northern Africa. Even their shops are right next to each other. I would be willing to predict that in less than ten years, this new Chinatown is going to surpass the old one in 13th District. The new Chinatown is filled primarily with people from Wenzhou, those intrepid souls well-known for venturing out into the world. I took a rough count that night, and there were close to ten stores with the words "Wenzhou" in their names.

Let's head back to the Latin Quarter, however.

Rue St Jacques, the main street of the Latin Quarter, was built along the remains of the city wall from the Roman period. It is also one of the oldest streets in Paris. To the west is the soul of the Latin Quarter – the Sorbonne. On the east side of the road and facing the Sorbonne is Collège de France, established by François I in the sixteenth century. At that time, the Reformation initiated by Martin Luther in opposition to the Roman Catholic Pope had just begun. It was also a time when the Society of Jesus was founded by Loyola, the Spanish soldier who saw it as his duty to defend the Pope in Rome. François I found the priests of the Sorbonne too conservative, which prompted him to establish a more liberal academy across the street.

At the entrance of the Collège de France is a Latin phrase *"Docet omnia"* (*We Teach Everything*), which may well represent the unique character of the Collège. During my visit, Professor Glowinski, the president of the Collège, and Professor Cohen-Tannoudji, a Nobel laureate in physics (currently a chair professor-at-large at City University of Hong Kong) invited me to lunch, where they briefed me on the Collège, and afterwards showed me around the campus. The Collège de France does not have any registered students, nor does it grant any degrees. It is primarily a community of scholars, and one has to be an authority in some area to teach there. The faculty is only expected to teach one course per year, but it has to be a new course every time. All courses are open to the public. Professor Glowinski has served as the president for over ten years. He is most proud of the substantial financial support he has managed to secure from two French presidents and several national assemblies, with which he has renovated the centuries-old buildings of the Collège, and constructed an information technology center and a large auditorium, both of which are underground (the only direction that any development can take given the status of the Collège as a historical site). The expansion and renovation process, lasting over ten years, was by itself valuable, but almost as important are the historical relics from the Roman period that were unearthed in the various stages of construction. To display these cultural relics, the Collège opened an archeological exhibition hall, which resembles a museum.

I am not being very precise with the name of the Sorbonne here. From the thirteenth century to the 1960s, the Sorbonne was regarded as a single entity. The rectangular structure on campus measuring several hundred meters long was built by Cardinal Richelieu in the seventeenth century when he was the prime minister to Louis XIII. During the French Revolution, when most of the clerics at the Sorbonne rallied in support of the monarchy and the church, the buildings were vandalized at the hands of a mob. They were renovated under the reign of Napoleon. In 1968, less than ten years after de Gaulle had established the Fifth Republic, university students in Paris staged a prolonged disturbance (the students called it a revolution; many of its leaders then have now become public dignitaries in France), forcing de Gaulle, who had twice stepped up to lead the nation in times of crisis, to withdraw, one year later when a referendum sponsored by him was defeated, from politics and retire to his country home. Later, the French government sought to expand higher education, and built a Université de Paris with thirteen independently administered campuses. The old Sorbonne was split into three universities, Paris III, Paris IV and Paris I, which I visited recently, in the southeast corner of the original

Sorbonne. Because its front gate looks across to the famous Panthéon, Paris I is also called Panthéon-Sorbonne.

The Panthéon is a symbol of the secularization (la laïcité) of the French Republic. It was originally meant as a church, but after the establishment of the Republic, the government stripped the building of religious meaning and turned it into a hall honoring distinguished French citizens. Statues of philosophers, men of letters, and scientists such as Voltaire, Rousseau, Hugo and Mme. Curie are kept there for public viewing. The power of induction into the Panthéon lies with the National Assembly (Assemblée Nationale). The most recent person to gain admittance to the Panthéon was the nineteenth century writer Alexandre Dumas Sr., who was the seventy-second person to be thus honored.

Behind the Panthéon is Saint-Etienne du Mont, a church whose architecture blends the Gothic style with that of the Renaissance. Most tourists turn back after they have seen the Panthéon, with the result that Saint-Etienne du Mont is rather quiet. Few know that the famous patron saint of Paris, Sainte Geneviève, is venerated in this church, and that Racine and Pascal, two literary giants, are buried there. Sometimes when I grew tired from walking, I would go inside for a rest, where I would ponder the ways of the ancients, and enjoy a tranquil moment.

On the two sides of this church are two famous secondary schools (Lycées), with impressive architecture to boot. They are Lycée Henri IV and Lycée Louis le Grand, from which many members of the French elite have emerged. The physical proximity of these schools to the most important universities in Paris has given their students an edge. Many of their graduates have gained admission to the Sorbonne, just a bit west across the street from them, or to Ecole Normale Supérieure, a bit to the south, Ecole Polytechnique to the east, or Ecole Nationale d'Administration further west. (Because of space limitations, Ecole Polytechnique and Ecole Nationale d'Administration have moved out of Paris.) Important scholars in letters, history and philosophy in France have by and large come from the first two of these institutions, while many engineers and scientists have graduated from Ecole Polytechnique, and prime ministers, ministers, members of the National Assembly, high-ranking civil servants, and chief executives of major corporations are alumni of Ecole Nationale d'Administration.

Until about twenty years ago, there were no private universities in France. All universities were directly under the Ministry of Education and were open to the public at no charge. In the past, all full professors in all universities were appointed by the Ministry; only in recent years have individual institutions been authorized to make appointments of their own. The best secondary school graduates do not necessarily regard entering universities as their first choice, unless they are accepted by major academic centers with long histories such as the Sorbonne. Rather, many are willing to devote one or two years to supplementary studies in order to get accepted by what are called *les grandes écoles,* such as Ecole Polytechnique, Ecole Normale Supérieure and others mentioned above. Many of these elite institutions were established during the Napoleonic era, and enjoy a higher status than universities.

In the last five days of my visit to Paris I, I moved from a three-star hotel near Cluny to a four-star hotel near St-Germain-des-Prés. To ease the pain of the increased expenses, I directed my thoughts to my "bitter" days of living on a daily budget of US$3.5 (including room and board) in 1963, and the "sweet" times of getting by on 350 euros nowadays (room but no board). In terms of location, décor, ambience and service, the hotel was impeccable. But who would have guessed that there was a lot to take care of from City University of Hong Kong in those few days! The editors from the *Academic Journal of Peking University* were also pressing hard for an article, with the result that I went into a marathon of writing and faxing the moment I got back to the hotel from the university each day. On a few occasions, the staff members on the night shift, whom I had gotten to know quite well, told me to take a break. There's always tomorrow, they said. They were probably wondering why anyone would squander precious euros to stay in such an expensive hotel, while all he ever did was spend all his time on work from Hong Kong. Would it not have made more sense just to go back to Hong Kong?

What they did not realize is that, as much as possible, I tried to go to a café of historical significance nearby for dinner and "work" every night. Le Procope, which I had visited a few times on my previous visits, is just a few steps away from the hotel. It was established in 1686, and claims to be the oldest café in France; Voltaire, Franklin and Napoleon were once regular customers. I would try to write as I ate. Even if nothing worth mentioning got written, well, at least I had the pleasure of pretending to be someone with refined taste. It is a pity that the waiters in our present corrupt age had manners that were far from perfect, often interrupting my thoughts with the bill.

Les Deux Magots (so named because of two wooden sculptures of Chinese trading agents there), a popular haunt for Hemingway and a group of surrealist artists, is another place worth a visit. Now, though, the only two things that one can do there are look and be looked at. Get out of Les Deux Magots, and go around the corner, and you will find yourself in front of Café de Flore, where Jean-Paul Sartre, the major existentialist thinker, and his companion, Simone de Beauvoir, held court with their followers. The ambience and décor there are quite fine, but my pen seemed to have come under the spell of Sartre's *Being and Nothingness* (L'Être et le Néant): there was no question about its being, but nothing would come out of it. All right, then, I thought, I'll go to Brasserie Lip just across the street and treat myself to Alsacian sausage and sauerkraut: the sourness of the sauerkraut might stimulate my thoughts. On one occasion, my thoughts were indeed stimulated and were just about to give off sparks when I noticed a message on my Blackberry telling me that the agenda of the next Senate meeting had to go out the following day.

Well, I had already made plans to go to visit the museum of Légion d'Honneur the next day, so I thought I'd better head back to my luxurious hotel to review the agenda.

In 1802, two years before he declared himself Emperor, Napoleon established Légion d'Honneur, which has remained the highest French order for two hundred years. Its award is the highest of all five national awards (such as Ordre National du Mérite and Ordre des Arts et des Lettres). The person presenting the honor is the commander-in-chief of the Legion, in other words, the President of France. A small red insignia from the Legion pinned to the left lapel of one's jacket commands respect and special regard from people in France. In recent years, many people from Hong Kong have expressed their friendship towards France and made significant contributions to the country. As a result, in proportion to its population, Hong Kong is probably one of the places with the highest number of awardees. Recently, the awardees of Hong Kong, numbering over ten, responded with great support to the call to renovate the museum of the Legion, and received special recognition from President Chirac, the commander-in-chief of the Legion. I feel deeply honored to be a member of this group.

Walk east along the left bank of the Seine from the museum of the Legion, and you will come to an old train station, which was almost torn down in the 1970s. It has since been turned into the Musée d'Orsay, which I had the opportunity to visit two weeks after it opened in 1986. As a teacher (in those days I carried with me my faculty ID from the University of Southern California), I even got in free by way of a special entrance. It stands to reason that as a country known for its promotion of culture, France should give special considerations to teachers.

Continue a bit further towards east, and a grand and splendid building comes into view. The building is called Institut de France, another organization that dates back to the Napoleonic years. Unlike the Collège de France, it is not a research and educational institution but the highest establishment of the humanities and sciences in the country. L'Académie Française, which holds the power to allow or disallow the inclusion of a word into the French dictionary, has its office there. Regulations stipulate that the academy cannot have more than forty members at any given time. The only, and historically the first, member of Asian ancestry is the translator, poet and novelist François Cheng (Cheng Baoyi), who came to France as a student from China at the age of seventeen. (The first Chinese intellectual who worked in France is a certain Huang from Fujian, who assisted in the compilation of the first Chinese-French dictionary at the court of Louis XIV.)

Also housed in Institut de France is L'Académie des Sciences. In May of this year, through the introduction of Professor Ciarlet, a member of L'Académie des Sciences and a colleague at City University of Hong Kong, I was given an invitation to attend a special celebration Séance Solennelle of the Académie. Honorary guards, clad in traditional military regalia, stood solemnly at the entrance of the venue. The Director, the President and the Permanent Secretary officiated at the celebration, all appearing in the traditional robes of the fellows. Some other fellows also wore robes, while more still came in suit and tie like other guests. The purpose of the celebration was to introduce the twenty-five new members, each of whom was asked to deliver a five-minute speech. Judging from their accomplishments and their speeches, there is no question that the French Republic still occupies an important place in various fields of knowledge in the world. France has its own complete technological system, and is at the forefront in areas such as the application of nuclear energy, air transportation, electronic communication, computer software and biotechnology. Its accomplishments are impressive beyond doubt. By virtue of achievements in the sciences alone, to say nothing of the wealth in literature, the arts, thought, and culture, France can not be easily dismissed by the facile label of "Old Europe" from the ill-informed.

A reception, with fine wine and food, came after the celebration. This was by no means the first time that I had partaken of a wonderful repast in L'Académie des Sciences. In 1999, I was invited by the French Foreign Ministry to visit the Académie. At that time, the President of the Académie, Professor Lions, treated our group to a sumptuous meal in a magnificent dining hall under a painted dome. Hemingway would never have guessed that it would be possible to have a meal like that in the solemn Institut de France on the left bank of the Seine.

Hemingway did, however, notice the magnificent residences on another island in the Siene, Ile St-Louis. I, too, have a special fondness for this quiet and lovely island, connected by a short bridge with Ile de la Cité. Apart from a few gift shops and distinctive inns, there are several restaurants which serve delicious food at prices that are not particularly exorbitant. In May of this year, my wife and I recommended this island to a Chinese mother and her daughter from New York. They could not imagine that in the hubbub of Paris, it is possible to find so quiet and comfortable a place as this. With gratitude, they even suggested treating the two of us to dinner.

Old Paris-hands that we are, we of course know of the famous La Tour d'Argent on the bank of the Seine, connected to Ile St-Louis by a bridge. Hemingway took note of it eighty years ago. He mentioned that at that time there were rooms for rent upstairs from the restaurant. Many visitors from Britain and the U.S. would leave their English books behind, which the restaurant kept in a small library. Given his financial circumstances at the time, my guess is that Hemingway probably did not eat there often.

For several decades, I have known that this is one of the best and most expensive restaurants in Paris, but it has never crossed my mind that I should go to try the food there. I cannot explain why, but thoughts of indulgence came to me this year, and I was planning to go there for a meal with my wife. My friends in France shot down the idea. Twenty-five years ago, Guide Michelin, the most authoritative food publication in France, awarded La Tour d'Argent three stars, the highest ranking. This year, it only merited one star. Disappointed, we nevertheless went with great enthusiasm to an inexpensive Tunisian restaurant, and filled ourselves with steaming couscous.

Many years have gone by. La Tour d'Argent is now but a faint shadow of its glorious past. Hemingway might never have tried couscous, but spring rolls, rice noodles and sushi are by no means strangers to the Parisian palate nowadays.

Paris, this wonderful movable feast, will continue to spread out and, like the Seine, flow forward with the waves of globalization.

Liberty, Equality, Fraternity

The French Revolution was an important event in human history. Its immediate cause was the storming of the prison at Bastille by the citizens of Paris on July 14, 1789, but its deeper roots are found in the social transformation and the promotion of rationalism and humanism in the two or three hundred years after the Renaissance. This was a revolution staged by members of the third estate of society (regardless of wealth) to topple the nobility who made up the first tier (represented by the monarchy) and the clerics who constituted the second tier (represented by the Church). The foundation on which the French Republic stands – liberté, égalité, fraternité – was proposed in 1793, the year Louis XVI was executed.

Today, most government offices and many schools in France have these three words engraved on them. This resounding declaration has inspired millions of people in France and throughout the world for more than two hundred years.

The course of human development is not without its ironies, however.

Just when the French Revolution had reached the point where fashionable people addressed each other as "Citizen So-and-so" in manifestation of the spirit of equality and fraternity, radical revolutionaries put together a Committee of

Public Security (Comité de Salut Public) to suppress all counter-revolutionary rhetoric and actions in various places. Guillotin, a physician by profession, was given the grave responsibility to study the most efficient method of executing the many who were sentenced to death. His invention ended up bearing his name – guillotine. What goes around comes around. In the end, many members of the Comité de Salut Public met their ends on the guillotine. Their radical actions and social disturbances led to restoration and counter-restoration, and the result was that Napoleon, who controlled the military, was elected by an overwhelming majority with almost no opposition to become the French emperor!

Civil law, units of measurement, government institutions, and the institutions of higher education mentioned above are all products of the Napoleonic era. They contributed greatly to the later development of France and even the world. Napoleon's adventurism (such as the invasion of Russia) and aggression against other European countries were quite a different matter, for which he ended up paying a heavy price. Perhaps it can be said that the French Revolution kicked off the progress of French industrialization, and in turn sped up the expansion of French colonialism. Napoleon was simply throwing himself wholeheartedly into the role given to him by his times.

"Liberté, égalité, fraternité," the banner of the French Revolution, also left an indelible mark on recent Chinese history. In May of 2006, I came to be a witness of this historical process.

In a place over one hundred kilometers south of Paris, there is a town called Montargis with a population of several tens of thousands. It is most famous for a kind of sweet called *prasline*, which is well-liked by the French (and every one in my family). Another reason for its fame is one of its former inhabitants, Anne-Louis Girodet, a well-known painter at the time of the French Revolution. That is why the museum of this small town has in its collection highly sought-after paintings that are not found even in the Louvre.

I went to visit Montargis not because of its sweets or paintings, however.

Between 1921 and 1926, the same time when Hemingway was staying in Paris, Deng Xiaoping and several hundred Chinese youths worked in a rubber factory called Hutchison near Montargis. In May of 2005, I was invited by a former mayor of the town to spend a weekend at his home, and got to understand in some detail the history of the work-study experience of Deng Xiaoping and others in Montargis. Only in their twenties and still highly malleable at the time, Deng Xiaoping, Li Fuchun, Cai Chang, He Changgong and other founders of the Chinese Communist movement lived as workers in this town and promoted revolution among Chinese living in France. Their actions might have drawn the notice of the French police, but they were never suppressed, and while their French ability could have been better, it is inconceivable that all the various schools of thought in post-World War I French society did not have an impact on their minds. "Liberté, égalité, fraternité" must have been a familiar saying that they took deeply to heart.

Had Zhou Enlai, Chen Yi, Deng Xiaoping and others not come into contact at such close range with the French Third Republic or the history of this nation after the French Revolution, what would have come of the development of China in the last half century, especially the period after the Cultural Revolution?

When I visited Montargis, I was part of another incident. Compared to momentous historical events, it is most insignificant but still interesting enough to warrant mention.

Isabey, my post-doctoral student, comes from Montargis. When he found out by chance that I had been invited to his hometown, he insisted I pay his elderly parents, whom I had met before, a visit. His father, now ninety-four years old, used to work on the railroad. His mother is eighty-seven years old. His family and my host had never met, but on account of this Chinese visitor who had come from a long way, they connected by telephone for the first time to discuss in which household I should dine for my meals there.

On Saturday morning, when my hostess was busy preparing a leg of venison in my honor, my host suggested he take me to the center of town to look at the weekly Saturday market. Two encounters of some dramatic import occurred there. First, we saw my student, baskets in arm, buying food in the market, in preparation for my visit at noon. I took it upon myself to introduce to each other the two families who had lived in Montargis for generations. We said goodbye after a short chat and walked to the other side of the market, where three men wearing T-shirts with the word "L'Humanité" printed on them were selling newspapers. My host dashed over to say hello, and introduced me to them. There was a man of seventy-five in the group, who, a member of the French Communist Party of long standing, had started out as a teacher and later became mayor of the city. The Communist Party was very influential at one point but is now in decline and *L'Humanité*, the party newspaper, is on the verge of bankruptcy. These three faithful party members were there trying to reverse the tide, knowing full well that they were attempting the impossible.

The decline of the French Communists has nothing to do with the disintegration of the Soviet bloc, but is rather a by-product of the long years when the Socialist Party was in power. Mitterrand was elected as the first Socialist president on May 10, 1981. I was in Paris then, and can still remember the wild celebration of Parisians on the streets and the blaring of horns throughout the night. After he assumed office, France raised the minimum wage, improved the public health system, and established a thirty-five-hour work week and thirty-five days of paid leave per year for all working individuals. Such measures crippled the French Communists, who had for a long time made no mention of class struggle but focused instead on social welfare.

France will conduct its presidential election again this May. This time, the left-wing Socialist Party put forward Ségolène Royal, a female candidate with relatively little political experience and without a sharp political identity. The right-wing party in power, Union pour un Mouvement Populaire (UMP), is represented by Nicolas Sarkozy, currently the Minister of the Interior with the image of a strong leader. Representing Union pour la Démocratie Française (UDF), François Bayrou, a former minister in right-wing governments, now runs on a middle-of-the-road platform and seems to enjoy good media and popular attention. The founder of the ultra-right-wing Front National, Jean-Marie Le Pen, is getting advanced in age, but persists despite repeated defeat. His power to rally support is not to be underestimated. Of late, there have appeared a number of other candidates with varying degrees of influence and recognition. It is still too early to tell who will win in the end.

As for what is on the minds of French voters, there are a few indications.

First, while most French people are nostalgic for the glorious days of the past, they have no intention to relive them. During the eighteenth and nineteenth centuries, French was the lingua franca in diplomatic circles, and French cuisine and manners set the standard for the entire world. After Peter the Great, although a follower of the Russian Orthodox Church, decided to push for westernization, France became a model for emulation among the upper crust of Russian society. Even Napoleon's invasion of Russia did not alter the course. The Russian nobility (including the Czar) spoke and wrote to each other in French. At the height of its power, Russia had designs on the Ottoman Empire, which, for its part, also modeled itself after France and, in an effort to strengthen itself against the Russians, employed French army officers to establish a military school and to train its army. Members of the upper class in the Ottoman Empire were just like their counterparts in Russia, and regarded it as a matter of pride to use French. As a matter of fact, the first official newspaper of the Ottoman Empire was printed in French!

Nowadays, although French is still the official language of a few dozen countries in the world, and the number of people who study French as their first foreign language is second only to that of English, French intellectuals are no longer unwilling or unable to use English in international conferences or academic journals as was the case twenty-five years ago. Today, French intellectuals are proud to be able to use English. While France will remain forever a strong nation in Europe and a wealthy and powerful country with extensive influence in the world in any scenario in the foreseeable future, most French set their minds somewhere else. A candidate who runs on a platform of restoring France to its former glory will not be elected President.

Second, even though France's economic growth has been rather slow and its unemployment rate remains high, on the whole, France is still an affluent and vital economic entity. Apart from a high level of technological development, France is able to make use of its former advantageous positions and wealthy cultural resources to manufacture many brand-name luxury products. For example, I ran into a handsome young couple at the Beijing Airport ten years ago. The husband had been sent to East Asia to be in charge of market exploration for Hermès. Today, the Asian market for Hermès products and those of other French brand names is comparable to the market for French airplanes, railroad system and nuclear power plant equipment. Ordinary French people live in material comfort, while the elderly and sick are adequately cared for. Even though many French scholars and commentators argue frequently as to whether France can maintain its economic prosperity and cultural uniqueness in the process of globalization, I personally am very optimistic. Neither do the ordinary French people seem to be overly concerned about it. A candidate who runs on a platform of globalization issues will therefore not likely be elected President, either.

In that case, what is uppermost in the minds of French people? What is the "deep-seated contradiction" of French society?

I can say categorically that what causes the biggest concern for the French population of over fifty million is the immigrants who come from north and west Africa, mostly of the Islamic faith, and their offspring, totaling more than five million. The riot in the fall of 2005 where cars were set on fire in many cities across the country has revealed to the fullest the existence of this conflict.

I purposely use the word "contradiction" rather than "problem" to describe this phenomenon because French society is complex and full of contradictions. The French Revolution and the subsequent republican systems (including the Fifth Republic of today) have emphasized the secular nature (*laïcité*) of political power. The word *laïcité* in French is slightly different in meaning from the English word secularism that is often used to translate it, for it underscores a principle stronger than that of the separation of church and state mandated in the American constitution. The French Revolution came under the influence of anti-clerical sentiment and, with the establishment of the Republic, the basic orientation of the polity was unsympathetic to the Catholic Church. Its initial ideal was to put the church (*l'église*) below the state (*l'état*); at one point, Catholic clerics were even put on the payroll of the state. During the Third Republic, a person who went to Mass on Sunday stood little chance of being selected for the cabinet.

The French Revolution and the American Revolution happened at around the same time, and their ideals of freedom, rule of law, and the independence of the three branches of government are very similar. Yet, the founders of the U.S. are believers in Protestantism. That the words "In God We Trust" are printed on their paper currency is enough to show that the American political entity celebrates rather than resists religion.

The complexity of French society lies precisely in its simultaneous distancing from and acceptance of the Catholic Church in the nineteenth century. While the country emphasized *laïcité* internally, French missionaries were actively propagating Catholicism around the world. The French government also regarded itself as the representative of the rights and interest of the Catholic Church. Today, the majority of French people identify themselves as Catholics, but Catholics who observe Mass on Sunday number no more than 15 percent of the population.

A strong and affluent France has acted like a magnet in history, drawing to it and assimilating successfully groups and groups of immigrants in the last two hundred years. There were the Italians, Spaniards, Portuguese, Polish, Irish,

Armenians and Romanians who were originally non-Catholics, and the Jews from north Africa and eastern Europe. And then there were the Protestant Germans from Alsace. This year's President-hopeful, Sarkozy, for example, is the son of Hungarian immigrants and his mother is Jewish.

In the mind of the average French person, the French Republic is the highest authority in the laiety, and citizens of the French Republic are equal and should not be divided by race, belief or ancestry. Religious belief is a personal matter, but the authority of the state should command respect. If you were to ask a "real" French person whether he or she regards himself or herself first as French or Catholic (or any other religion), the answer you would get would always be the former. But, were you to ask the same question of the immigrants (or their off-spring) from Algeria, Moroco or Senegal, the answer might very well be different.

In the last fifty years, a large number of Muslim immigrants from former French colonies in North Africa and West Africa have come into the country. In general, they come from a lower educational background and economic status. In their religion and cultural traditions, the concept of the separation of church and state does not exist. Neither do they appreciate the meaning of *laïcité*. The overwhelming majority of them are engaged in low-income jobs, and since the social welfare system in France supports them when they are unemployed, they are definitely unwilling to return to their place of origin. Through the free education provided by the government, their children gain a command of the French language and identify themselves as French citizens. At the same time, however, they also feel that they are not given the treatment due the French. Looking for jobs is not easy for them, either, and they are filled with feelings of frustration and grievance.

In C.E. 732, Charles Martel, grandfather of Charlemagne, repelled the invading Muslim forces from north Africa who crossed the Pyrenees in Poitiers in western France, thereby establishing a country that belonged to the Franks. Today, the Muslims do not invade by force, but are attracted by the better job prospects and social welfare system in the home of their former colonial master. There are no signs as yet that they will be assimilated in any great numbers. On the one hand, the French government respects their freedom of religion, and on the other, fails to come up with a feasible proposal to incorporate the huge numbers of Muslims with no tradition of the separation of church and state into the fold of mainstream French society that upholds the principles of *laïcité*.

This is the biggest social contradiction in France since the French Revolution. This is what today's French voters are most concerned about.

Marie Curie, Albert Schweitzer, Claude Lévi-Strauss

The seventy-two eminent persons honored in the Panthéon are those who have won the formal recognition of the state for their accomplishments. As a matter of fact, since the eighteenth century, there have been numerous other individuals who are worthy of a place in history.

Even when I was in high school, I had great admiration for Marie Sklodowska-Curie; as an undergraduate, I held Albert Schweitzer in highest regard; and from my American roommate when I was a graduate student, I heard about the anthropologist, Claude Lévi-Strauss.

Mme. Curie and her husband were professors at the Sorbonne and won world fame for their discovery of radioactive chemical elements. She was the first woman in the world to win the Nobel Prize, and she won it twice, first in physics in 1903 and then in chemistry in 1911. Paris VI is named Pierre et Marie Curie after her and her husband.

Schweitzer was a famous organist, musicologist, church pastor, surgeon, and winner of the Nobel Peace Prize in 1953. When he was a pastor in his hometown, he made the decision to go on a mission to Africa where he also hoped to practice medicine. He spent the next eight years studying medicine, and earned a medical doctorate degree in 1913. In the same year, he opened a hospital in the French colony Gabon in Africa. In the half century that followed, apart from six years when he lived in Europe and the frequent trips that he took to Europe and the U.S. to raise money for his hospital by giving speeches and organ performances, he spent all of his time in Gabon until his death in 1965.

Lévi-Strauss studied philosophy when he was a student at the Sorbonne, but became an explorer and an anthropologist of the tribal people in the forest along the Amazon when he was sent to teach sociology at the University of Sao Paulo in Brazil and to conduct field studies of the Brazilian aborigines. He taught in New York during World War II. After the war, he returned to France and drew the world's attention with his *Tristes Tropiques*, written in a moving and fluid style. In 1959, he was appointed the chair professor of anthropology at Collège de France. The book *Structural Anthropology*, which established his position in the field of anthropology, was also published at this time. The most important work in his life is probably *La Pensée Sauvage* published in 1962, in which he raised objections to the existentialist thinker, Jean-Paul Sartre. (The title in French is a double-entendre, meaning "Savage

Mind" and "Wild Pansy" at the same time.) Lévi-Strauss is now ninety-nine years old.

Mme. Curie, Schweitzer and Lévi-Strauss, highly respected intellectuals in France, might not have known each other, but there is a common point among them. They were all minorities born outside of France. Mme. Curie was Polish, born in Warsaw. As a tribute to her motherland, she even named the radioactive element she discovered polonium. Schweitzer was born in Strasburg, the capital of Alsace which France ceded to Germany after the Franco-Prussian War. He was German-Alsacian by heritage, raised in a Protestant family and educated in Berlin when he was young. After World War I, Alsace was returned to France, whereby he became a French citizen. Lévi-Strauss was Jewish, born in Brussels, Belgium. After World War II broke out, he returned from Brazil to join the resistance forces against Germany, but with the establishment of the pro-Nazi Vichy government, he could not but leave France for New York.

That these three distinguished individuals enjoy such high esteem from society for their accomplishments is due partly to their talents and efforts, but also partly to the openness and inclusiveness of French society. These are the principles of "liberté, égalité, fraternité" and laïcité in action. If one were to point out that, for all of their differences in religion, these three individuals were all white Europeans, and hence the openness of French society is rather limited in scope, then it could also be argued that the admission of François Cheng, an Asian immigrant, into the Académie Française represents the further development of French inclusiveness.

When we examine the biggest social issue of today's French society from this angle, we begin to see glimpses of a solution.

At present, young people of north or west African origin

who are accepted into the best universities and institutes (*les grandes écoles*) are far and few between. Even those who enter the average institutions of higher education are rather limited in number. Consequently, unschooled and un-employed, offspring of these immigrants further find themselves living in poor conditions. It is only natural that they harbor feelings of alienation and resentment.

In the January/February issue of this year's *Foreign Affairs,* the publication of the Council on Foreign Relations in the U.S., is an article by Dominique Moïsi, a senior advisor of Institut Français des Relations Internationales, entitled "The Clash of Emotions." Moïsi argues that in today's world, "the Western world displays a culture of fear, the Arab and Muslim worlds are trapped in a culture of humiliation, and much of Asia displays a culture of hope." In examining the state of mind of the Muslims in France (his name suggests that he may be of Jewish descent), he believes that "the riots that took place during the fall of 2005 had an essentially socioeconomic origin, but they were also a lashing out by the disaffected against a society that claims to give them equal rights in principle but fails to do so in practice."

This is an observation worth pondering over. Do most French nationals of European descent see this point or do they have some apprehension about a very distinct minority that is increasing in number? What could allow them to shake off the "culture of fear" as they face the future? On the other hand, without concrete and real examples of achievement and success, it is probably difficult for French citizens of African and Arab origins to eliminate their sense of frustration and humiliation.

As an outsider who cares about France, I wonder whether it is possible to find a common way to dissolve these two different emotions? Would full implementation of the foundation of the French Revolution two hundred years ago, "Liberté, Egalité, Fraternité.", not be the solution? To promote ethnic and racial harmony within its boundary, and to face the challenges of globalization, France needs to transform its citizens of north African and west African origins, who make up ten percent of the population, into competitive and hopeful French nationals.

One can only resort to higher education to realize this goal. If the French government could institute a policy of incentives (something similar to the affirmative action policies in the U.S., or the practice of giving bonus points to ethnic minority university applicants in China), it would definitely speed up the process of having more immigrant children in the institutions of higher learning. When they graduate, they can join the rest of the white-collar workforce in jobs with prospects. This will be an important step in dissolving the contradictions in French society.

Education is a slow process. Although it is a good solution to the existing contradictions in the French society, it cannot be accomplished in just a short time. Even if the French government is determined to come up with a policy of incentives, it will take time for its implementation. More time is needed for this policy to achieve initial success and become widely accepted.

This will take at least one generation, perhaps take even two or three generations. During this process, the attitude of the families and the cultural orientation of the communities of the Muslim population will play a key role in its success or failure. At the moment the majority of Muslim families in France has several children and the parents are in general not well educated themselves, thus unable to help the children with their school work or not putting enough emphasis on their children's education.

The content of a secular education provided by the French government may very well conflict with the value system that the pupils acquired from their homes and communities, creating unavoidable dilemmas or confusion for the Muslim pupils. Under such circumstances, dropping out of school may seem a reasonable choice. To such youths in academic difficulty and/or mental anguish, the counseling and encouragement of the parents and elders in the community is very important.

Muslims in general possess a strong faith in divine revelation and many believe in predestination. Some scholars therefore think that this may lead some Muslims, particularly those who are not well educated, into a tendency of fatalism, thus showing less initiative in changing their lives and possibly having an impact on acquiring the competitiveness that is needed in the contemporary society. Although these socio-cultural issues are extremely complex, I am nevertheless optimistic about the future of the Muslim population in France. In the 1960s, I observed at close range the Civil Rights Movement in the United States, and have since witnessed the definite improvement of racial relations in the U.S. after the passage of the Civil Rights Act by the American Congress and a very discernible rise in social status of the Alfro-Americans. Who in those days could have foreseen that America's first two Secretaries of State in the twenty-first century are both Afro-Americans, who, as a group, were struggling for equal opportunity in education only forty years ago?

Suppose that, in the years to come, a group of highly respected scientists, writers and literary figures, sociologists, and political and corporate leaders of Arabic and African origins emerges: then, the value systems and feelings towards France among the French Muslim population will change. In turn, the concept that a government secular in nature can nevertheless treat religion with respect will spread among them. Such a development will exert a great inspirational force (just like the French Revolution) to the Muslim population in other Western European countries, or even Islamic countries in their efforts at modernization.

I have faith in the wisdom of the French people. Two hundred years ago, they came up with the stirring slogan of "Liberté, Egalité, Fraternité," and fifty years ago, with great resolve, they gave up their colonial empire and reconciled with their sworn enemy, Germany. Together, the two countries created a new face for Europe. In the future, the French people will realize the ideal of the founding of their country and use it to resolve the social contradictions of today.

When a large number of young Muslims study at the Sorbonne, Ecole Nationale d'Administration, and Ecole Polytechnique, when they can take as their models such people as Mme. Curie, Schweitzer and Lévi-Strauss who were born outside of France and brought up in different religions, the two groups of French citizens will emerge from their emotions of fear and humiliation and march towards hope.

(March 2007)

3

Thoughts by the Ganges

After New Year this year, I went on an academic visit to the University of Delhi in India. This was my third visit to India. I went there for the first time in 1963. My ignorance then only allowed me to appreciate things at their most superficial; I was like a dragonfly skipping over the surface of the water. I made my second visit in 2001 at the invitation of the Indian Council of Cultural Relations. For all my enthusiasm, I was still only able to gain a cursory glimpse of India. This time, though, I stayed in New Delhi for an entire month. I came into contact with many people, and had the opportunity to experience life in India.

For my first two weeks at the University of Delhi, I stayed in the university's guesthouse for international scholars. I took all my meals at the dining hall of the guesthouse for more than ten days. The food was fine and the price reasonable, but the menu remained unchanged day after day: roti or rice, dahl, paneer, mixed vegetables, stewed lamb or stewed chicken.

The fact that lamb and chicken took turns appearing on the menu has social roots: Hindus do not eat beef, and Muslims do not eat pork. That these two communities have lived with and adapted to each other in the last several hundred years accounts for the absence of beef and pork in most restaurants in India.

I moved to a hotel in the latter half of my month in New Delhi. At night, I could often hear the racket of wedding banquets at the hotel. Most of these wedding celebrations were upper-middle class affairs. The ladies always dressed in colorful, eye-catching clothes, and the splendor and ostentation of these occasions matched that of comparable celebrations held in the mainland of China nowadays. In India, however, the guests, together with the families of the bride and groom, often form parade lines, singing, dancing, and making the ceremonies far more lively and romantic. Indian classical literature is full of richly imaginative and fantastic stories, while our Confucian classics, such as *Analects* and *Mencius,* contain nothing more than starchy moral instructions about the regulation of the self and the restoration of ancient rites. Do these two different sets of cultural texts affect the character and habits of their people? From my own observations of wedding customs, the answer is assuredly yes.

Let Them Eat Lamb and Chicken,
and More of It

To be sure, India is a marvelously vibrant country, and deserves to be called a living human museum of multi-cultural life. The country has more than three hundred dialects, with no less than fifteen official languages, of which eleven belong to the Indo-European language family, and four to the Dravidian family. Even though Hindi, the language in use in the heartland of India, was made the national language at the time of India's independence in 1947, most Indians with higher education still use English in many situations.

In terms of skin color and facial features, the Indian population is made up of three groups: the Dravidians living mainly in the south, who have dark skin and low-ridged noses;

the Aryans living mainly in the northwest, who have light skin color and high-ridged noses; and the Tibetan-Burmese living mainly in the northeast, with their yellow skin tone and medium-ridged noses. In reality, most Indians are descendants of the union of the former two groups. Within the same region, town or village, or even within the same family, one can find Indians of different phenotypic features.

Language and skin tone are external differences. As far as the social fabric is concerned, the most fundamental criterion that Indians use to differentiate themselves from one another is religion. Within Hinduism, moreover, the distinction by caste plays an even more important role.

About 82 percent of the more than a billion people who make up the Indian population are observers of various sects of Hinduism. Another 14 percent believe in Islam, and a considerable number believe in Christianity, Sikhism, Jainism or Zoroastrianism. The biggest crisis lurking in Indian politics and society today lies in the conflicts between the Hindu and the Muslim communities.

After the eleventh century C.E., Persian-speaking Turkish Muslims from Afghanistan invaded India in stages and established many regional political powers. Gradually, they became the rulers of India, except for the southernmost regions. The most powerful among them was the Delhi Sultanate with its capital in Delhi. In the mid-sixteenth century, another Persian-speaking group of Turkish-Mongolian origin invaded India and established the Mughal (the Indian mispronunciation of the word Mongol) Dynasty that was to last for three hundred years. Its founder, Babur (1483–1530), was descended from the lineage of Timur and Genghis Khan. After Babur, there came to the throne a number of kings equally adept in domestic, diplomatic and military affairs, which explains why this small group of

Persian-speaking Muslims was able to rule over such a wide territory inhabited by so many more Hindus. Yet, for all that, the last thousand years of Indian history has seen India, with all of its small principalities, in a state of feudal division.

During the eight hundred years of the Muslim reign, a large number of Indians, especially royalty and Hindus of the lower castes, converted to Islam. This is why in different places in India today, there are people of different languages, skin color and social status among the believers of Hinduism and Islam.

Apart from their advanced technology and organizational skill, the British were able to govern India partly because they were able to exploit the divisions within Indian society and the presence of these numerous smaller states.

After World War II, Britain, powerless to suppress the independence movement of the Indians any longer, was preparing to withdraw from India. The Indian National Congress made up mostly of Hindus and the Muslim League were two of the political entities that had fought for independence for many years. The former proposed to establish a secular state made up of different ethnicities and religions throughout all of British India. The latter demanded the establishment of a separate Islamic country comprising the territories occupied by a clear majority of Muslims. Many bloody conflicts erupted between these religious communities in 1946–1947, prompting Britain to grant independence to India ahead of schedule. Under the maneuvers of Lord Mountbatten, the uncle of the reigning Queen Elizabeth, the Indian National Congress and the Muslim League arrived at an agreement establishing separate rule, and authorized Britain to draw the boundary between what became India and Pakistan. In legal terms, there were more than five hundred independent states in India at that time. The ruler of the states, called Maharajahs, could decide which of the two newly established countries to join. Kashmir, located in the northernmost part of India and Pakistan, had a predominantly Muslim population, but the maharajah was a Hindu, who remained indecisive for a long time. After religious conflicts erupted, however, he decided to join India, which triggered the first war between India and Pakistan. Fifty years later, Kashmir is still divided, with half governed by India and the other half by Pakistan. The problem has yet to be resolved.

In the early years of Indian and Pakistani independence, one million Hindus and Muslims died in religious strifes, and more than ten million fled their homes in order to resettle in the newly established states of India or Pakistan. From this perspective alone, it has been no easy task for India to have persisted in its policy of religious diversity and maintained relative social peace for all these years. Nonetheless, the situation remains tense among the various religions.

The primary purpose of my visit to the University of Delhi was to establish contacts with scholars who specialize in the history of Indian civilization and study its relationship with other civilizations. At the invitation of an advanced research institute, I participated in a seminar with a group of scholars. When I explained the plan of my visit to those present at the seminar, two professors said to me, "What do you mean by Indian civilization? It is something we are still arguing about."

Indeed, it is. Some people believe that Indian civilization has at its core the Vedas compiled three thousand five hundred years ago when the Aryans established themselves in South Asia. According to this view, the Indus Valley civilization discovered by archeologists in modern times dates to the untraceable ancient period, whereas Islamic culture was an external culture brought in by invaders during the Middle

Ages. Another view maintains that Indian civilization is composite in nature, its primary elements being the lifestyle and the social order of Hinduism such as the caste system, and its secondary elements, more prominent in later periods, being those of Islamic culture, evidenced by such features as the style of architecture and male attire. Of course, Sikh and Christian civilizations should be included as well.

Without exceptions, all the Indian Muslim scholars I know support the latter view, but both views have supporters among the Hindu scholars I know.

As a matter of fact, the term "Hindu civilization" is only a vague notion. Even the word Hinduism itself was first employed by Westerners to refer to the religion subscribed to by most people who lived in Hindustan or India. The reason that this term was acceptable to the Indians is that most Hindus define their identity by the caste system prevalent in the region of their birth. There is no single deity, scripture or religious leader who is recognized by all, a feature that distinguishes Hinduism clearly from Islam or Christianity.

At one point in the past, Shakyamuni broke away from Brahmanism to establish Buddhism. When it declined in India in the seventh century C.E., Shakyamuni was looked upon by many Indians as a deity in the Hindu pantheon. In essence, then, Buddhism was absorbed into Hinduism and vanished from India.

From the past to the present, many people have believed that Hinduism and Islam (at least its Sufi sects) could be integrated with each other because they are similar in their practice of meditation and the repeated chanting of prayers and their transcendental experience of the union of the man and god. The outstanding ruler Akbar (1542–1605) of the Mughal dynasty had made exactly this kind of attempt to bring the two religions together.

In reality, the dream of integrating Islam with Hinduism will not work. When I was in India, the cricket tournament between India and Pakistan was in progress. India lost the first game, and the Hindus told me right away that many Indian Muslims were in fact rooting for the Pakistani team. At the same time, there were Muslims who came out to the open to declare that they of course supported the Indian team. This indicates to me that there are still ill feelings between these two social-religious communities.

After I came back from India, I came across two items of news. In the first story, Varanasi Ghats, by the side of which several millions came to bathe in the river on festive occasions, had been vandalized by terrorists. In the second piece of news, a series of explosions had disrupted the gathering of Muslims on a Friday at Jama Masjid, India's biggest mosque not far from the University of Delhi.

The best scenario that I can imagine for these two communities is for them to coexist peacefully at normal times, and to negotiate with each other when problems arise. If these two major communities can lead their lives without eyeing each other with hostility or disrupting each other's lives, it would not be too bad, either. On the other hand, if they are led by the nose by extremists, with the result that their relationship remains tense all the time and conflicts break out every now and then, then it would be unsettling indeed. If one day the friction becomes such that it leads to a large-scale confrontation, then it would be a big tragedy for South Asians, and yet another demonstration of the weakness of human nature.

I am not pessimistic, but I dare not be optimistic either. I only hope that in the future, I can continue to eat lamb and chicken in India, and more of it.

The 1857 Incident and More

In the thousands of years of Indian history, no other incident has aroused more comments from history scholars and political figures than an incident that broke out in 1857, and no other incident has received more textual documentation than that same incident in 1857. The British called it a "mutiny"; Indian officials call it the "First War of Independence"; historians in general call it an "uprising," while some scholars in the West call it a "revolt" or "rebellion." I shall call it "the 1857 Incident" here.

In the spring of 1857, soldiers employed by the British East India Company were given a new kind of rifle and bullets that required them to bite off the cartridges with their teeth before they could load them into their guns. Some people believed that the lubrication under the cartridges was made of animal grease, which caused extreme fear and dissatisfaction among the Indian soldiers (called Sepoys). The Hindus among them thought that there was cow grease in the lubrication, and hence a sacrilege to their gods, while the Muslims soldiers believed it was pork lard, which was an insult to Islam. Beginning from Calcutta, the headquarters of the British East India Company, some Indian soldiers refused to use this kind of new ammunition. Rumors and acts of disobedience soon spread to other places where troops were stationed.

At the time, eighty-five of the ninety soldiers in a cavalry company stationed not far away from Delhi who refused to use the bullets were court-martialed. They were subjected to public humiliation, and then were given sentences of imprisonment of eight to ten years. The next day, many Indian soldiers forced their way into the prison, set their colleagues free, and killed the officers and the other British nationals there. Three days later, these and other Indian soldiers took Delhi, and with the hope of setting up a symbol of resistance to drive away the British, they installed on the throne once again an old Mughal emperor, who, at eighty-two, was without soldiers and power. The old emperor made tours around the city on top of an elephant, inciting people to the cause of resistance, which spread like prairie fire to the various places along the Ganges valley. It spread as far as the south, where it attracted people of different classes to its ranks. Many Hindu princes and kings who had either been removed from their offices or had their power weakened also participated. Most Indians, however, merely watched from the sidelines, and there were some, most notably the Sikhs, who came actively to the aid of the British soldiers. The whole incident went on for close to a year. Several thousand British nationals were killed in different places by Indian soldiers and civilians. On one occasion, the British soldiers were on the verge of recovering a city on the Ganges plain, and the Indians decided to execute the British in their custody. The soldiers who had received military training were unwilling to fire on women and children, and instead, the people picked a few butchers among themselves to kill over two hundred British in the way they would butcher animals.

The most startling episode of the whole 1857 Incident took place in a place called Lucknow in the center of the Ganges Plain. Since the decline of the Mughal dynasty in the eighteenth century, Delhi had not been in a position to command the rest of India. Some members of the nobility continued to govern a number of Muslim states under the title of Nawab (similar to governor). Among them, the most important is the Kingdom of Oudh, with its capital in Lucknow. In 1856, the British took over Oudh, doing away with the privileges of the feudal lords and thus alienating many members of the upper class.

Oudh was a wealthy place, well-known for the jewelry and embroidery produced there. It is also famous for its cuisine, which boasts a kind of mutton roll that is delicious and tender, and seems to melt in one's mouth. In the eyes of Indians in general, people from Oudh are cultured and have good manners. In order to appreciate the ethos of the place, I took advantage of a weekend to fly to Lucknow from Delhi, where, in addition to indulging myself in the sights and food, I also had a chance to learn a historical lesson in a most vivid way.

The classroom of this lesson was at a high point outside of the city of Lucknow, the site of the residence of the British representative that was burnt down in 1857.

This "Residency", now in ruins, covers a big area. There are many buildings, with an aura similar to the Yuan Ming Palace that the allied forces of the British and French burned down three years later. At that time, the British Empire installed a "Resident" in various states in India. In order to underscore the importance of such a post, the British constructed in many places Western-style buildings, which served as the seat of authority and were called Residencies. After the 1857 Incident broke out, Indian officials and civilians killed many Christian missionaries, and the safety of other British nationals came under threat. Several thousand British from neighboring regions gathered at the Residency at Lucknow. With the advantage of its high location and the ammunition and food gathered there, they were prepared to stage a forceful resistance. For five months, even under the barrage of heavy canon fire, the Residency remained standing. This was reported with high approbation in British newspapers at the time, and the dead (numbering approximately two thousand, some from cannon fire and some from cholera and dysentery) were eulogized as war heroes. Later, reinforcements came in from Persia, Ceylon, Malaya and other places and recovered Lucknow and the few buildings that remained of the Residency.

The British forces were filled with profound hatred at the brutality of the Indians. Upon the recovery of Lucknow, they retaliated just as they had done when they reclaimed Delhi, where they let loose a spree of murder, rape and plunder.

After the dust had settled, the British exiled the old Mughal emperor to Burma, where he was imprisoned, and declared that India would come under the direct rule of Queen Victoria, with the responsibilities of governing India transferred from the British East India Company to the Secretary of State for India in the cabinet. The newly appointed Viceroy reorganized the British military stationed in India, and targeted for new recruits the ethnic groups as well as the region from which they came that had remained loyal to the British during the Incident. Moreover, the ratio of British to Indian soldiers was expanded from one to six to one to three. Britain also consciously limited the activities of missionaries, and did not continue to attempt to convert Indians to Christianity.

Even though Christianity was no longer thrust down the throats of the Indians, the British authorities persisted in their policy of educating the Indians in English, because it was based on the proposal of Lord Macaulay (1800–1859), who had argued some twenty years previously: "We must at present do our best to form a class who may be interpreters between us and the millions whom we govern; a class of persons, Indian in blood and color, but English in taste, in opinions, in morals, and in intellect."

When I was in high school in Taiwan, the English

grammar book that I used was written for Indian students during the era of British India. Now fifty years later, when I compose this essay, my mind is filled with thoughts. Had the British not persisted in teaching English to the Indians, would there have been so many Indians whose English writing receives wide acclaim in the English-speaking world? Furthermore, had India not come under British rule, would there be a unified Republic of India today?

Logically paradoxal that they are, I cannot come up with a good answer to these questions, but it is clear that the Indian government today comes across as being full of confidence. First, in the literature about the museum of the Lucknow Residency and the Memorial of the Incident, the 1857 Incident is now referred to as the "First War of Independence," and the Indians who died at that time are honored as martyrs. Secondly, the Indian government is now devoting its energies to cultivate contracts with transnational corporations to process their commercial corporations, and to participate in the operations of the global economy, through which they hope to accelerate the modernization of India.

This seems to be the logic behind their thinking: the 1857 Incident was a righteous movement in which the un-enlightened and backward Indians of the nineteenth century resisted British colonialism. To affirm on the whole the resistance movement (including some of its most barbarous actions) is not to diminish the dignity of the Republic of India as a modern state. All the more, it does not mean that India should give up its continued quest for modernity in the twenty-first century.

For those scholars who study the history of modern China and look forward to seeing the modernization of China, the 1857 Incident of India can perhaps be a case study for their reference.

"Factory of the World"
versus "Office of the World"

People who have had any contact with Indians will find that Indians are quick with thoughts and words, articulate, and good with numbers. My Indians friends, however, told me that these are the characteristics of only certain social classes or castes. Most people from the poor peasantry and laboring masses are not like that at all.

Nevertheless, India nowadays has a prominent advantage in the overall framework of global economy: namely, it has a massive low-cost human resource made up people who are properly educated and proficient in English. With the help of telecommunication and the Internet, the big companies and organizations of the rich countries can transfer to India for processing those services that do not require face-to-face interactions, especially those that are related to commercial operations. (Such a phenomenon has given the English language a new word, "offshoring".) Last year, about six hundred thousand Indians were engaged in providing services to businesses and other kinds of enterprises, from security control of buildings to accounting, from medical tests to software development. These services generated for India about US$24 billion in foreign exchange, averaging a per capita productivity of US$40,000.

Most Americans are getting used to hearing English with a slight Indian accent these days. It does not matter whether one wants to ask for directory assistance, discuss credit card records, reserve a plane ticket, or solve a computer installation problem, as long as one is looking for service on the phone, half of the time the service providers are Indians.

Of course, most consumers are not immediately conscious of the presence of Indians in higher-ranking jobs such as engineering design or radiology diagnosis.

Because of its rich human resources, India is going to enlarge her role as the "Office of the World." Relative to China's role as the "Factory of the World," which is feared and castigated by Europeans and Americans, India is in a much better position. On the average, one Indian can generate US$40,000 worth of products. Just imagine how many factory workers, and how many pairs of shoes it would take to generate the same value in China. And India can do all this without harming its environment or depleting its natural resources, while having in its hands the various kinds of data from the people, commercial establishments and even governments of many countries.

Many people in Indian society are of course supportive of its role as the "Office of the World." However, about 80 percent of the people of India are engaged in agricultural work, and 40 percent of them are illiterate. The role, therefore, can only be played by an extreme minority. For most Indian farmers and minor commercial and industrial workers in the cities, it is still too early to say whether the impact of the global economy will do them good or harm. In terms of purchasing power parity, India's per capita income is only one-tenth that of the United States, and half as much as China's. There are over three hundred million people with a daily actual income of less than one American dollar. Considered from the point of view of her demographic structure and social economy, it will be a very long time before India can pull herself out of poverty.

China's role as the "Factory of the World" has not been free from controversy at home. This is why the Chinese authorities these days emphasize creativity and innovation, with the hope that China can come up with new products and services and developing brand names of her own. I openly elaborated on this concept at a symposium organized by the Ministry of Education two years ago for presidents of Chinese and overseas universities. But it is one thing to grasp the idea, quite another to put it into practice. Chinese tradition has all along stressed the importance of showing obedience due to the older, the higher and the collective. Parents require their children to be docile, teachers require their students to listen and follow orders. How can we expect to change overnight this habit of mind that has been in existence for one thousand years?

Creativity and obedience are two distinct states of mind. Creativity requires motivation and a special ability that comes from within. It is absolutely not true that there is no creativity in Chinese. The poetry of Li Bai and Du Fu, the movable type of Bi Sheng, or even the use of paper currency and bank draft are manifestations of high creativity. But the geniuses of Li Bai and Du Fu did not come into being due to the conscious promotion of Emperor Xuan Zong of Tang. For that matter, the creativity of Thomas Edison and Albert Einstein in the modern U.S. or Europe was not the outcome of instructions from their parents. But living in European and American society, these two families were at least ready to defy the norm by allowing their children to discontinue schooling. Suppose the parents of these two geniuses had insisted that their children listen and follow orders; it would be very difficult to say what kind of accomplishment we could have expected from them. Turning to the current situation in China, it will also take a long time for her to shed her role as the "Factory of World" and to increase her wealth by means of creativity and innovation.

Today, China and India are the two developing countries

that are giving rise to a lot of discussion in the West. There are many similarities between these two ancient countries with a combined population of 40 percent of that of the whole world, and there are many complementary areas on which the two can collaborate.

In history, Buddhism imported from India has altered the spiritual life and esthetic sensibilities of the Chinese. This is where Indian civilization has brought about fundamental change in China. Because of war and successive changes in religions, materials about the most glorious period in Indian history, the Buddhist dynasty of Gupta, can no longer be found in India. Today, Indian historians derive their understanding of this period mostly from the records in Fa Xian's *A Record of Buddhistic Countries* and Xuan Zang's *A Journey to India in the Great Tang Dynasty*. China can thus repay its cultural debt to India. At the same time, this is an example of the mutual benefit the two countries enjoy from friendly cooperation.

From Zhang Qian's excursion to Central Asia in the Former Han dynasty to the British incursion into Tibet towards the end of the Qing dynasty, India and China have always been at peace with each other. At the beginning of the twentieth century, Britain unilaterally declared the McMahon Line on a portion of the border between India and China, and in 1947 transferred this border definition to the newly independent India. Today, the dispute over the border between the two countries has not yet been resolved. At present, China and India are two poor developing countries, each with a large territory; there is still a huge distance to cover before they can claim to be truly wealthy. They should turn their attention to developing their economies and promoting social harmony. How they can cooperate with and complement each other will be a matter of concern not only for these two countries but for the whole world.

Fifty years ago, on the eve of India's independence, Jawaharlal Nehru said in an exclusive interview with a reporter from the Central News Agency stationed in New Delhi that Asia's future would be assured if India and China would hold together. When he uttered those words, Toynbee, the British historian, had yet to make his prediction that the twenty-first century would belong to Asians. My wish is that the leaders of these two countries will be broadminded enough to entertain such a vision.

(June 2006)

4
Thoughts Along the Bosporus Strait

"There is No History Without Geography"

A fundamental premise runs through the thought of the French historian Fernand Braudel, "There is no history without geography."

The first time I took the ferry across the Bosporus, the waters were calm, but emotions surged like waves inside me. The strait spans no more than eight hundred meters wide here, but it separates Asia and Europe even as it serves as the meeting point of Eastern and Western cultures. Much later, I took a cruise from the Sea of Marmara that connects with the Mediterranean on the south, through the Bosporus, to the Black Sea. Along the way, I saw cliffs on the two banks of Asia and Europe, the clumps of trees that grew on their surface, and the luxurious mansions among them. All I could think of was the rise and fall of various empires over the past two thousand years.

With the precipitous crags on its two shores, the Bosporus indeed separated the Balkan Peninsula from Asia Minor until modern times. It is claimed that no ancient army ever crossed this narrow passageway, be it the invading force of Darius on its way to Greece, or that of Alexander the Great on its way to Persia. Even though it is no more than thirty kilometers long, the strait is the hub of commerce between the Black Sea and the Mediterranean, and the choke-point that controls the traffic between them. In the Russo-Japanese War of 1904,

Russia's Far East Fleet was defeated by the Japanese navy, and it became necessary for the Russians to dispatch their ships from the Black Sea to join the battle in Asia. A Japanese intelligence agent, who had up to that point worked for thirteen years to promote Japanese goods in Istanbul, the capital of the Ottoman Empire, hired some people to keep an eye around the clock on the movements of Russian ships across the Bosporus Strait from the vantage point of the Galata Tower.

The two shores of this significant historical waterway belong to one city – Istanbul.

Because its geography boasts the Golden Horn, the best natural harbor in the world, as well as a strategic position that makes it highly invulnerable to invasion, Istanbul has remained both a prosperous commercial center and a political capital for two thousand years. Its earlier name was Byzantium, well-known for its maritime traffic as early as the era of the Greek city-states. In 313 C.E., the Roman emperor Constantine moved his capital to Byzantium, and the name was changed to Constantinople, which it remained until the demise of the Byzantine (East Roman) Empire in 1453. For more than a thousand years, it was the most glorious and developed city of the Christian world.

In October of last year, I was in Turkey for three weeks of academic visits, one of which was spent in Istanbul. It happened that during that week, my friend Orhan Pamuk was awarded the 2006 Nobel Prize in literature – the first time that a Turkish writer had received such an honor. I thus had the opportunity to observe at close range the different responses of the people in Istanbul to this piece of news.

At the end of November, Pope Benedict XVI, who earlier had offended Muslims world-wide in a speech, came to visit Turkey and called upon Christians and Muslims to come together for dialogue. During his visit, he prayed side by side with the Grand Mufti of Istanbul in the famous Blue Mosque (so named because the interior of the mosque is decorated with blue tiles). This was the first time in history that the leader of the Catholic Church had prayed in a mosque. In addition, the Pope held mass with the Patriarch of the Occidental Orthodox Church, where the two joined in giving benediction to the congregation.

On December 8, Pamuk delivered his Nobel Prize lecture, entitled, "My Father's Suitcase," at the Nobel Prize Award Ceremony in Sweden. Contrary to the wishes of many people, he made no mention of the massacre of Armenians at the hands of Turks or the human rights situation in Turkey today.

In late January of this year, Hrant Dink, a Turkish reporter of Armenian descent, was murdered in Istanbul. A few days later, the police arrested a seventeen-year-old youth suspected of the murder and a number of alleged abettors. Close to 100,000 people attended the funeral of this reporter in Istanbul, including the Deputy Prime Minister, the Minister of the Interior, and many other dignitaries of Turkey.

Even though Istanbul and the Bosporus are extremely important geographically and historically, they would not have made it to the pages of Hong Kong newspapers if not for these news stories. Similarly, if not for the war in Bosnia, the civil strife in Kosovo, the disturbances in Macedonia, the conflicts in Cyprus, or the Israeli-Palestinian clashes, the turmoil in Lebanon, and most recently, the Iraq War, many people would not have known so much about the territories of the Ottoman Empire of the past, and the many problems that came after its dissolution. Perhaps I may be allowed to rephrase Braudel, "There is no news without history. And without news, we will never learn so much about geography and history."

Eternal Istanbul

಄಄಄಄಄಄಄಄಄಄಄಄಄಄಄಄಄಄಄಄಄಄

Legend has it that when the troops of the Ottoman Empire were advancing on Istanbul, they asked local peasants for directions. The peasants pointed at the domes of the churches and the tall city walls in the distance and declared in Greek, "*Is tan polin!*" (*To the city!*) In the ears of the Ottoman troops, the city thus gained a new name, Istanbul.

Of the many Turkic groups that invaded Asia Minor (also known as Anatolia) of the Byzantine Empire after the eleventh century, the Ottomans advanced the furthest. Although they had been "Persianized" to a large degree over several centuries as they moved from Central Asia to Anatolia, accepting part of Byzantine culture, they still retained the tradition of Turkic-speaking tribes after their conversion to Islam, and regarded themselves as warriors (*gazis*) of its vanguard. Taking advantage of the weakened state of the Seljuk Turks who ruled the area at the time, Osman, the first leader of the Ottoman Empire, established his own princely-state in the western part of Anatolia in 1299. His son, Orhan, began the conquest of the Balkan Peninsula. In 1453, when the 21-year-old Mehmed II led his troops to lay siege to the isolated Constantinople, the last stand of the Byzantine Empire, his army included the mounted troops led by Serbian Orthodox Christians who had earlier surrendered to the Ottomans.

In the critical moments of the siege of Constantinople, the Genoan and Venetian merchants in the city reached into their own purses to request troops from western Europe to come to their rescue. Bent on fighting the Ottomans, the ninety-fifth and last Byzantine emperor went so far as to express a wish to be reconciled with the Latin Church with the hope that it would help him fend off the advance of the non-Christians. The memory of the pillage of Constantinople at the hands of the Fourth Crusade, whose leaders also attempted to establish a Latin Kingdom in the region, lived on, however, and the clergy of the Greek Orthodox Church preferred to surrender and live with religious autonomy under the rule of the Islamic sultan rather than take orders from the Latin Pope.

This leads us to an important question: how did the Ottomans govern their vast territories and the ethnically, linguistically and religiously diverse groups under their rule?

The answer is the *"millet* system", established in accordance with the practices of previous Islamic rulers. The House of the Ottomans regarded itself as the protector of the Islamic faith, and categorized its subjects into "believers" (the Muslims) and the "non-believers" (the non-Muslims). The latter was further divided into those who followed the teaching of a Holy book, "peoples of the Book," as they were called, and those who did not. "People of the book" referred to those who adhered to the teachings of either the Judaic scriptures or the Christian Bible. The Ottomans established *"millets,"* or religious communities, among these different groups of people in different parts of the empire. Muslims were first-class citizens, whereas believers of other religions had to pay a poll tax, but matters of religion, education, marriage and inheritance were the prerogative of the individual *millet*, which enjoyed a high degree of autonomy. People from different millets dressed differently and were easily distinguishable from one another. Ordinarily, there was little communication among *millets*.

The Ottomans, therefore, divided people under their rule by religion in administering the empire, and a person's race or ethnic origin was not important at all. Even though the Ottoman royal family originally came from the Turkic groups in Central Asia, they had inter-married often with other races in the several hundred years of their migration. After the establishment of the empire, many queens and *odalisques* were Slavs, Circacians, Arabs, or even Africans by origin. The ethnic make-up of the royal family was rather complicated as a result. Many *grand viziers* and high officials were not Turkic people; some were even Christians or members of the Jewish faith.

Until the beginning of the nineteenth century, the Ottomans also practiced what was called the *devşirme* system, whereby young Christian boys (many of whom had been captured in war) considered to possess special physical and intellectual aptitude were chosen to serve as personal slaves to the sultans. They were given the best education, including the teachings of Islam, and when they grew up, they were appointed officials, palace aides, or members of the janissary corps according to their ability. Sinan, the most distinguished architect of the Ottoman Empire, and many other officials, were products of the *devşirme* system.

By the eighteenth century, Istanbul had grown almost to the scale that it is today. Contemporary statistics showed that there were 2,000 large mosques, 6,000 small mosques, 670 courts of law, 2,000 schools, and 19 hospitals. In addition, there were 997 caravansarays, 657 small inns, 989 water pipes, and 9,995 drinking fountains. Among residential quarters, 990 were Muslim, 354 Greek Orthodox, 657 Jewish, 27 Armenian Christian, and 17 "Frank," that is, European Catholic. (Under the pressure of the British, French and Russians, the Ottoman Empire established new *millets* of various denominations of the Christian faith in the nineteenth century.)

Such is the Istanbul that I saw: a modern, multi-ethnic, multi-cultural cosmopolitan center that has inherited the two major cultural streams of Persia and Byzantium. With two thousand years of cultural sedimentation, the city has assumed a beautiful and enchanting exterior and a graceful and substantive interior.

Istanbul competed for the right to host the 2008 Olympics, but eventually lost to Beijing. No one can deny, however, the city that straddles the two continents of Asia and Europe, where Eastern and Western cultures converge, is the eternal Istanbul.

"The Sick Man of Europe"

All of the first ten sultans of the Ottoman dynasty were competent in warfare and diligent in court affairs. Much longer than the prosperous reigns of the Kangxi, Yongzheng and Qianlong Emperors in China combined, this period of the Ottoman Empire is unmatched by any dynasty in human history. By the time of the tenth ruler, Suleyman, the "Magnificent" , the "Lawgiver" (r. 1522–1566), what started off as the Osman Emirate had become a colossal empire that spanned the three continents of Europe, Asia and Africa. It stretched from the Adriatic Sea to the Persian Gulf, and from Egypt to the Caucasus. The three Islamic holy cities – Mecca, Medina, and Jerusalem – came under its rule, and Hungary and Crimea became its tributary states. In 1529, Suleyman laid siege to Vienna. It is said that when the Ottoman army retreated, they left behind sacks and sacks of coffee, which marked the beginning of the habit of drinking coffee among the Europeans. We will put aside the question as to whether this story is apocryphal or not, but history clearly shows that at that time European craftsmen often traveled to Istanbul to learn the skills of making better clocks. In moments of internal strife, European rulers would also dispatch envoys bearing gifts to the Ottomans to ask for help.

It is a historical pattern that a country that has enjoyed a period of prosperity will eventually come to decline. Such is the case with the Ottoman Empire after the death of Suleyman, first because of disputes over succession of sultans, and secondly due to the corruption of officials that robbed the empire of its vibrancy. Thanks to the Enlightenment and maritime activities, the social and economic development of Europe was at a rapid pace over the same period, and the balance of power shifted as a result. The turning point of the Ottoman Empire, with its history of over six hundred years, came in 1683 with the defeat of its second incursion into Austria. The Treaty of Karlowitz of 1699, was then forced upon the Ottomans by the European powers, leading to the ceding of Hungary, Croatia, Slovenia, and other places.

Overall, it is nothing short of a record in human history that the Ottoman family managed to maintain its rule for more than six hundred years. The first two hundred and fifty years of its long history was a period of expansion. Its territories grew, as did its population, and the sultans had enough human resources to deploy and enough land to reward the officials for their military exploits. The middle hundred years can be considered a period of change. The territories had stopped growing, and the methods of production fell behind those in Europe. With the discovery of new sea routes which the Europeans took advantage of to travel to Asia, income from commercial activities also declined. At the same time, the sultans, having grown up in the protective environment of the palace, were unable or unwilling to lead troops into the battle, while members of the royal family and their favorite followers were given rewards that they did nothing to deserve. For the Ottoman Empire, which in essence was a military state, such changes portended its eventual collapse. The question that remained was how long the period of decline would last. Surprisingly, it lasted more than two hundred years. In these two hundred years until the Empire came to an official end in 1922, from the Turks (not what they called themselves, but a name given to them by the Europeans, just as the Ottomans called all Europeans "Franks"), the mere mention of whose name was enough to inspire fear among Europeans, they became the "Sick Man of Europe," which was only invoked with contempt.

The "Sick Man of Europe" had very little contact with the "Sick Man of East Asia" under the Qing rule of China, but they resembled each other in many ways. First, as a result of military defeats that they suffered, both were forced to cede land and pay indemnities. (The enemies of the Ottomans were Russia, Austria, Venice, and later France and Britain.) Secondly, both countries had to cede control over their own tariffs, without the protection of which local commerce and industry did not have a chance to develop, and benefits obtained by one foreign country were quickly demanded by other powers. The third similarity lay in consular juris-dictional rights. In the eyes of the Europeans, the laws of the Ottoman Empire were backward, and the European powers therefore demanded to set up courts of law of their own.

The Ottoman Empire was after all part of Europe; at least, it was not far from the "nucleus of Europe," unlike Qing China, whose emperors and officials basically knew nothing about Europe. In the eighteenth century, the Ottomans began to learn from the military technology and education systems of the West. A top-down reform campaign, termed *Tanzimat* (Restructuring) that lasted some forty years officially began in 1839, and went further than the "Westernization Movement" in China, sending students to West Europe, establishing new styles of schools, opening a translation bureau, disbanding the old janissary corps, and inviting European military advisors to train a new army. Later, a postal service was started, as was telegraph service and train service. The empire reformed governmental work practices and increased its efficiency, and non-Muslims were allowed to participate in politics on an equal footing. In the end, these reforms paved the way for constitutional monarchy. The draft of a constitution was passed by the parliament in 1878 but was suspended for thirty years by Abdulhamid II, a contemporary of Emperor Guangxu, whose role resembled that of Guangxu and the Empress Dowager Cixi rolled into one. By the time he agreed to adopt constitutional government in 1908 due to internal pressure from elite groups, the Ottoman Empire was in its last gasps, like a sun about to set behind the horizon.

The Ottoman Empire and the Republic of Turkey

People familiar with the modern history of China cannot but be struck by the strong resemblances between the "Sick Man of Europe" (the Ottoman Empire) and the "Sick Man of East Asia" (China). Yet, on further consideration, rather than seeing these resemblances as indicative of any commonalities between the Ottoman Empire and the Qing Empire, it may be more appropriate to regard them as special characteristics of European imperialism. Think of Iran, Afghanistan, the Mughal Empire in India and Thailand – did they not all go through similar experiences at the hands of the European imperialists?

As a matter of fact, there is a big difference between the

Ottoman Empire and the Qing Empire, which in turn shaped the fate that awaited them. Both the Republic of China and the People's Republic of China consider themselves the successor to the Qing Empire: the name of the country and the system of government might have changed, but not its territories or its people. But today's Republic of Turkey was established in Anatolia and Thrace in the southeastern corner of Europe through a succession of battles. Since the nineteenth century, after European nationalism made its way to the Ottoman Empire, the Serbians, Greeks, Romanians, Bulgarians, and Armenians formed independent states at various points with the help of the European powers, while the regions inhabited by Arabs (including Iraq, Syria, Jordan and Palestine) became protectorates of the British and French after World War I.

At this time, the pragmatic among the Ottoman elite accepted what history and reality had laid down for them. They gave up the practice of the Ottomans of categorizing people not by race or ethnicity but by religious allegiance, with the hope of galvanizing what strength there was left in the empire to form a modern republic made up of primarily "Turkish" people in Anatolia where Turkic-speaking groups had lived for close to a thousand years. The founder of this republic was Mustafa Kemal Atatürk, a blue-eyed, Turkish-speaking Muslim born in Salonika (in today's Greece). After the establishment of the Republic of Turkey, Atatürk propounded in a series of lectures on the "Six Principles" of the Republic, known in the West as "Kemalism." (This happened at almost the same time as Sun Yat-sen's lectures on "The Three Peoples' Principles" in Guangzhou. The two are in fact similar in nature.) The tenets of Kemalism spell out the need for Turkey to modernize through Westernization, to secularize its society, to recognize religious faith as a personal matter, and to grant equal status to all citizens. Thereupon, Turkey did away with the Arabic script in its Ottoman-Turkish language, but adopted Latin letters in writing. People were no longer allowed to wear *fez* (a kind of tall hat), and every citizen had to adopt a surname. (The Turkish Grand National Assembly passed a law that gave Kemal his last name, Atatürk, which means the father of the Turks.)

Mustafa Kemal Atatürk was a modern soldier who graduated from a military academy. He was also a fervent Turkish nationalist who admired Westerners for their accomplishments but did not identify completely with the West. His views on parliamentary democracy might have influenced Chiang Kai-shek's insistence on "from military government to political tutelage, and from political tutelage to constitutional government.", a theory first enunciated by Sun Yat-sen. Indeed, the Turkish military has always been the loyal defender and executor of Kemalism. Whenever an Islamist political party rises in influence and power, the military has stepped in to intervene. (The current Prime Minister Erdogan was put in jail at one point for reciting in public an ancient poem that had a militant Islamist tone.) Neither has the military by any means been gentle in its treatment of liberal intellectuals who identify themselves with the West. (Orhan Pamuk was indicted for violating a law that forbids actions injurious to the dignity of the Turkish people. The case was later dismissed on technicality). In order to join the European Union, the Turkish army has had to make appropriate concessions in the last few years. Before then, however, the job of defending the constitution invariably fell to the National Security Council in which the military wielded strong influence over the civilian ministers by running its secretariat. In fact, in the period after World War II, the military interfered with civil affairs in the name of protecting the constitution every ten to twenty years, before handing political authority back to the people or the politicians. From this point of view, one could say that the European Union has forced the Republic of Turkey to move from the stage of "political tutelage" to that of "constitutional government."

It is still difficult to determine where this change will take Turkey.

Pan-Islamism and Pan-Turkism

On the whole, history seems to show that before an era of great changes, elite groups with social resources at their command become polarized. For self preservation, some advocate change, while others do their best to maintain the status quo. Their conservative or progressive tendencies lead them to behave differently, follow different schools of thought, and march under different slogans, such as the movements to "reform under the guise of the old" in China and "uphold the monarchy and fend off the barbarians" in Japan.

As the Ottoman Empire was being devoured piece by piece by the European powers, almost to the brink of collapse, two kinds of thinking emerged from the Ottoman elites that would prove influential to the international scene in the future. The point of departure for the first, "Pan-Islamism," was the revival of Islam, while the second, "Pan-Turkism," bolstered its cause by appealing to "ethnic feelings."

For several hundred years, the Sultan of the Ottoman Empire proclaimed himself as the *khalifa*, the highest leader in the Islamic world. "Pan-Islamism," with an agenda to restore the former glory of Islam and its emphasis on a community (*umma*) to which all Muslims in the world belong, had a strong appeal to the upper layers of Ottoman society, who regarded Islam as a unifying force to bring together all Muslim peoples and hoped to use this as a force to resist the Christian Europeans.

Today, the Pan-Islamic sentiment found in other countries (such as the Muslim Brotherhood in Egypt) can hardly be said to have come directly from the Ottoman era, but such thinking has an undeniable appeal for Muslims living under the harassment and persecution of non-believers. Yet, persuasive as it may be in theory, Pan-Islamism has been rather ineffectual as a weapon when pitched against actual conflicts over power and interests. During World War I, under the instigation of the British and the French, the Arabs on the Arabian Peninsula rose up against Ottoman rule, dealing a serious blow to Pan-Islamism in the homeland of Islam. The Kurds living today in the southeastern area of Turkey and northern Iraq are Sunni Muslims, but this has not stopped many of them from engaging in separatist activities.

Under the sway of European nationalism since the nineteenth century, various ethnic groups living in the territory of the Ottoman Empire have sought to establish independent countries of their own. The Turkish- (or Turkic-) speaking Muslims were inspired to unite all "Turkic people" of the world. To these Muslims, the label "Turkic people" included all speakers of Turkic languages in history. They went so far as to designate a place of origin for all Turkic people in Turan, the whereabouts of which are unknown. This is simply wishful thinking.

Indeed, the Turks originated from a place north of today's Mongolia. The Turkic languages and those of Xiongnu, Xianbei, Kitan, Jurgen, Mongols, Manchus and Koreans all belong to the Altaic language family. Beginning in the sixth century, the Turkic people migrated westward along the steppes of northern Asia and inter-married with different groups along the way, absorbing the languages of those they conquered. Today, from Turkey to Xinjiang in China, the languages spoken by and large belong to the same group of languages, the "Turkic languages." However, speakers of these languages are widely disbursed geographically and are made up of people of distinct ethnicities. Furthermore, their historical experiences are all different. To conclude from their common linguistic origin that they are the same people is far from scientific. As Pan-Slavic thinking gained ascendancy during the nineteenth century in Russia, some elites in the Ottoman Empire also harbored illusions of Pan-Turkism, hoping that this would unite speakers of Turkic languages under Russian rule, such as the people of Azerbaijan, Turkmenistan, Uzbekistan, Kyrgyzstan, and Kazakhstan. In reality, Pan-Turkism did not produce any real effect east of the Caspian Sea. The disintegration of the Soviet Bloc in recent years saw the independence of republics made up primarily of speakers of Turkic languages, giving rise once again to a new Pan-Turkic thinking among some people in Turkey. The European Union's recent decision to reject Turkey but extend a welcoming hand to Bulgaria not only represented a setback to the wish entertained by the upper crust of Turkish society for a hundred years to "withdraw from Asia and join Europe," but also dealt a serious blow to Turkey's self-esteem. It would not be at all surprising if some Turkish people might want to re-enter Asia and be the leader of Turkic-speaking countries. According to my own observations, however, most people in modern Turkey (including those with Islamist inclinations) have set their goal at joining the European Union rather than seeing themselves as part of Asia.

To be sure, a Turkish government of any political orientation will place considerable weight on the cultural and economic relationship it enjoys with other Turkic-speaking countries, given the close connections in their language and religion, and the abundance of oil and natural gas in all of them. This is particularly the case with Azerbaijan, a part of the former Ottoman Empire, where the language is almost identical to modern Turkish. However, this is absolutely not the same as promoting "Pan-Turkism." Some "East Turkistan" elements within China fantasize that Turkey, on account of "Pan-Turkism," may lend them support in their separatist movement. This kind of thinking is completely unrealistic.

The Turkish March

The *Turkish March*, one of Mozart's piano sonatas, is a world famous piece of music. Where his inspiration came from, and why he named this piece of music as he did, I have no idea, but we can be sure that the ancestors of today's inhabitants of Turkey indeed "marched" from both East and West to where they are now.

As early as 4,000 ago, there already existed in Anatolia the Hattian civilization established by the Semites from the southeast. About 3,800 years ago, a west Indo-European-speaking tribe came into the same area and established the strong Hittite civilization. About 3,300 yeas ago, another invading force, the Indo-European Phrygians, came from Europe. They were later replaced by the ancient Greeks, who were themselves replaced by the Persians during some periods. Even now, remnants of ancient Greek civilization and of the Hellenic period can be found in Turkey. Not long before the birth of Jesus, the Romans conquered this piece of land. Caesar's famous line, "I came, I saw, I conquered," (*Veni, vidi, vici*) was uttered to the Roman Senate as he moved on today's Turkey. For this reason, many traces of Roman civilization and the sites of early Christian activities are also found in Turkey.

When the Turkic tribes marched to Anatolia in the eleventh century, therefore, the place already boasted a continuous civilization of three thousand years, including one thousand years of Roman rule and then Byzantine Empire. Several hundred more years of military action and inter-marriage went by and the primary language in Anatolia changed from Greek to Turkish.

It can be seen that when the Republic of Turkey decided to "withdraw from Asia and join Europe" in the twentieth century and apply for admission to the European Union, there was indeed a historical, cultural and ethnographical basis for it.

Nevertheless, the main difference between Turkey and other European countries (with the exceptions of Albania and Bosnia, each with a small population) is religion, which, many Turkish people believe, is the reason that they have not been admitted into the European Union. One cannot of course rule out such a factor, but given its expansive territories (equivalent in size to Germany, the largest country in European Union), its large population (almost the same as Germany with the highest population in European Union) and its relatively undeveloped economy (behind Greece), even if it were not a Muslim country, its participation in the European Union would produce an impact not to be underestimated.

The irony is that the more the European Union drags its feet in accepting Turkey out of a sense of insecurity, the more influential Islamism will become in Turkey, which of course will make Europe feel even more insecure.

Although the clites in Turkey are doing their best to

downplay the role of religion and to erase unpleasant memories of the past, under the current tense situation in the Middle East, especially since the Iraq War, many Turkish people have become more fervent in their Islamic faith, and nostalgic about the Ottoman era.

Eighty years ago, the newly established Republic of Turkey launched a campaign similar to the New Culture Movement, emblematized by the "movement to dismantle the Confucian establishment" and the "vernacular movement," that took place at about the same time in China. Today, intellectuals and the middle class in Turkish cities (especially those in the western part of the country) on the whole are inclined to identify themselves with the West, and many have also accepted the value system of modern-day Europe. They are no longer able to read Arabic writings in the mosque or the language used in the Ottoman period; neither can they read what is written on the many tourist souvenirs sold in the shops.

I visited five universities in Turkey. The way that the teachers and students dress is no different from what I have seen in the universities in Europe. In direct contrast to Iran, Saudi Arabia and other places, but similar to France, women students are not allowed to wear head scarfs or veils by law! (No such legal restrictions are found in countries such as the U.S., the U.K., Germany, Egypt, Jordan and Palestine.)

Obviously, secularism has taken root in Turkey, and the cultural leaning towards the West is quite evident. I could not find more than a handful of bookstores that sold foreign books in the whole city of Cairo, not even in the vicinity of Cairo University, but in the cities of Turkey where I visited, one can find English publications at many places.

This, however, is only one side of Turkey. In this country that straddles Europe and Asia, where Eastern and Western cultures converge, there is another face, which can be seen in the eastern regions of the country. The peasants in the villages and members of the lower social class are by no means Europeanized. Even in Istanbul, the most Europeanized city, the chant of the *muezzin* is broadcast at high volume from the top of the minarets of mosques five times a day to remind people that Turkey is still a Muslim country. In the narrow alleys and in any bazaar, tourists can still witness traces of history that cannot be easily erased.

Just as in China, where the Royal Feast of Complete Manchu and Han Courses has come back into fashion, where televised historical dramas on the Qing dynasty have enjoyed wide popularity, and where many people advocate the revival of Confucianism, including the study of the Confucian classics among young students, in Turkey today, many restaurants featuring a menu of Ottoman cuisine are drawing in large crowds, and books and television programs about Ottoman history and culture are increasing in number.

When I visited Turkey last year, it was Ramadan – the month of fasting for Muslims. In Bilkent University in Ankara and Bogazici University in Istanbul, I saw a lot of people eating lunch in the university dining hall, which goes to show that many Turkish intellectuals no longer observe fasting, one of the five basic precepts of Islam. At the same time, in the area around Istiklal Cadessi, the busiest and most Westernized street in Istanbul, there are a number of restaurants serving traditional Ottoman cuisine, most beloved by the local middle class. At night, when Muslims are allowed to eat during Ramadan, these restaurants are filled to the brim, illustrating yet another face of Turkish society.

Clothing and food are the most explicit manifestations of a culture. My observations vividly demonstrate that the path to modernization is a circuitous, complex and difficult one. It is no easy task to shake off history, in Turkey as much as in China.

There is a big difference between China and Turkey, however. With China's geographical location, history and the situation that it is in today, there is only one option open to it: to become a great modern nation in Asia by revitalizing its people and civilization. Lying ahead of Turkey, on the other hand, are three paths. First, join the European Union and become a modern, democratic and secular "light green" Muslim country in Europe. Second, stay away from the European Union and form a "light green" Muslim regional cooperation sphere with countries that used to be part of the former Ottoman Empire, such as Azerbaijan, Syria, Lebanon, Iraq, Jordan, Palestine and possibly Egypt. Third, turn "dark green" from "light green," and form a Middle East Islamic alliance that includes large Sunni Muslim countries such as Saudi Arabia and Egypt, which might also turn "dark green" due to its own internal forces, or even the major "dark green" Shiite country, Iran. This move would create a major shock in Europe.

The first path is the choice of the contemporary Turkish government and the vast majority of its people. The second is a choice that many people would reluctantly pursue only if Turkey is finally rejected by Europe. The third is one that radical Islamists would hope to see and the result towards which they would work to realize.

Two hundred years ago, Mozart composed his *Turkish March* in Vienna. In the future, Europe may well decide in Brussels the direction in which Turkey will march.

(April 2007)

5

From *The Adventures of Hajji Baba* to *My Name Is Red*: A Personal Journey

A Rendezvous in Istanbul

One day after a blistering snow storm hit Istanbul in January 2004, when the sidewalks were still mostly covered with snow and the outdoor temperature still below zero, my wife and I took a stroll along the posh Istiklal Cadessi (Independence Avenue) in the Beyoglu section of the city north of the famed Golden Horn. Having just seen the British Consulate building that was badly damaged not long ago by a terrorists' bomb, feeling both cold and hungry, we chanced upon a restaurant, Cafe Hajj Baba,[1] named after the main character of a nineteenth-century book that I had read some twenty years earlier.

We had a simple but pleasant lunch in the Cafe, but I did more than eating. With the help of the Turkish Consulate-General in Hong Kong, I had arranged to meet the famous Turkish author Orhan Pamuk during our five-day visit to Istanbul. Pamuk was traveling in India when the Turkish Foreign Ministry tracked him down and informed him of my request. Equipped with his telephone numbers, we arrived in Istanbul only to catch the worst snow storm in years. While we had to reshuffle some of our planned visits, we were more frustrated by our inability to connect with Mr. Pamuk. No one knew where he was or when he was to return. Now we had only two days left in Istanbul, meeting him became even a higher priority; his latest book *Istanbul*, was just then a hot topic on radio and television. Also, the Provost of the famous Bogazici (Bosphorus) University, Professor Sevket Pamuk, whom we met as the snowstorm was gathering strength, turned out to be Orhan Pamuk's brother.

Out of desperation, or perhaps inspiration, during lunch, I thought of sending Orhan Pamuk a fax message. The manager of Cafe Hajj Baba complied with my unusual request because "Mr. Pamuk is a good customer of our restaurant."

The next morning, one day before our departure, I received a call from Orhan Pamuk; he had just returned to Istanbul the night before and seen my fax. He said he would meet us for dinner at the Divan Hotel, not far from Cafe Hajj Baba.

We had no problem spotting Mr. Pamuk in the restaurant, for he looked just like his picture on the back cover of *My Name Is Red*, which I had read with great interest and admiration. Over an otherwise ordinary dinner, we talked about many subjects, mostly on the art of miniature painting, around which the plot of his book *My Name Is Red* unfolded. But we also discussed such topics as the future of Islamic societies and the mood swing of many Turkish people in the past decade, particularly after the war in Iraq broke out. All this had a somewhat surreal air about it.

I, as a biomedical engineer, became motivated to visit Istanbul not because I had read much about Constantinople in the Byzantine period or Istanbul in the modern time; nor because I had been an amateur of the history and culture of Islamic societies. In the summer of 2003, just before we left Hong Kong for a family reunion, a friend, knowing my long-standing interest in Islam, gave me a novel set in the old city of Istanbul. In the one week on a cruise ship, I read this intricate, delightful and vividly written book twice. I was so absorbed in this book and attended so few other activities on board the luxurious cruise ship that I was dubbed "My Name is Red" by my children and other relatives who found me antisocial at times. It was during that cruise that my wife and I decided to spend our Chinese New Year holidays of January 2004 in Istanbul.

Tales of the Alhambra

I first met a Muslim when I was a little boy in Jinan, Shandong Province. I had a neighbor and classmate, who seemed to live exactly the same way as I lived except that he never ate anything in our house. His family, he explained, were Muslims (*hui min* 回民). Later in Taipei where I grew up,

I watched first the construction and then a kind of majestic lonely existence of a mosque next to a very busy Catholic church. I visited the church numerous times, but somehow never once had the curiosity to peek into the mosque.

From Taipei I set out to study at Stanford University in the San Francisco Bay Area in July 1963. For the purely technical reason of obtaining my student visa where my parents lived, I had to fly to Ethiopia, where my parents were working for the World Health Organization (WHO), via Hong Kong, Bangkok, Beirut and Khartoum; the latter two were capitals of countries with a majority of Muslims. Later in my life, I had several occasions to meet with some Muslims, but I did not feel particularly interested in their history and culture till 1982, when I was in Paris on a one-year sabbatical leave from McGill University.

In January of that year, I took the chance to visit Spain, the land of Isabella and Ferdinand, Loyola, Cervantes, Velázquez, Goya, Picasso, Dali and Casals. Impressed by Spain's colorful history, rich folklore and the thousands of castles dotting its landscape, I, somewhat curiously, became enamored with the Islamic influence on Spanish architecture, music, dance, and the other arts. Just as I first learned about the twentieth-century Spain through the pen of an American author, Earnest Hemingway, I first learned of the Islamic past of Spain through the fanciful tales told by an earlier American author, Washington Irving.

The Tales of Alhambra, first published in 1832, is a delightful book to read, particularly if you have traveled, as I have done, from Sevilla to Granada. For the first time since I heard the term Muslim when I was seven years old, my intellectual curiosity was aroused by what I saw and read about in the eight hundred years of Muslim (*Moorish*) presence in the Iberian Peninsula, which the Moors called *Al-*

Andalus. The mosques, palaces, gardens were so exquisite, and the tales about their construction and destruction so fantastic, that I almost instantly turned into an *aficionado* about the various *taifas* (small kingdoms) and successive *emirs* (chieftains) that had Cordova, Sevilla and Granada as the political, commercial and cultural centers until 1492.

Even though Spain lay in the western fringe of the once expansive and powerful Islamic Empire, it was the most advanced and enlightened part of medieval Europe. Muslims, Christians and Jews lived in peace and worked together to produce agricultural outputs and handicrafts that surpassed in quality and variety those from other places in Europe. The scholars among them translated Arabic and Greek works into Latin for all Europe to read, particularly in the twelfth and thirteenth centuries when Spain became the centre of learning and a beacon of light in Europe's search for the way out of the Vale of Tears, a term used by the medieval Catholic Church to describe this world of ours.

"To God belongs the East and the West"

In the year of 622 of the Common Era (C.E.), Muhammad made his famous *Hejira* (emigration) to Medina from his birthplace Mecca, where he had achieved only very moderate success in winning converts to the new faith as revealed to him by *Allah (God)*. When he died ten years later, the strictly monotheistic religion he had founded had not only been accepted by most tribes in the Arabian Peninsula, but had also forged a new Arab state with Medina as the source of its spiritual and political power. Motivated by the new faith and strengthened by the new state organization, the Arab people set out on a path of conquest and established, within a hundred years of Muhammad's death, a vast empire spreading out from Spain in the west to the borders of China in the east. This outburst of energy and sudden expansion was unmatched in history and was to many Muslims the testimony of what is said in the Koran: "To God belongs the East and the West."

Although Arabic became the common language of this vast empire, especially in religion and state affairs, and numerous conquered non-Arab peoples did adopt the Arabic language, many local languages and cultures also continued to flourish and provided a great diversity within the empire. During the Umayyad Caliphate (661–750 C.E.) with its capital in Damascus, many Hellenized Arabs and Greeks helped the Arab Muslim rulers with administration as well as the construction of a new civilization. The most numerous and prominent group of non-Arab converts to Islam were the Persians. Inhabiting the land stretching from Mesopotamia in the southwest to Transoxania (east of the Aral Sea between Cyr River and Amu River) in the northeast, the Persian-speaking people with their ancient culture exerted a strong influence on the development of Islamic polity, philosophy, literature and arts. Because of their familiarity with statecraft, the Persians often occupied the most important positions in the Abbasid Caliphate (750–1258 C.E.) centered in Baghdad. Due in large measure to the Persian contribution, the ninth and tenth centuries were generally considered the Golden Age of the Islamic culture.

Towards the end of the tenth century, Persian nationalism within the Arab Empire began to take an overt form of expression when Firdausi wrote the book *Shanameh* (*Book of Kings*) as an account of Persian history, and Nizami wrote his five epic poems, *Quintet,* which have become the popular legends in the Muslim cultural tradition. The love stories of *Khusrau and Shirin* and *Layla and Majnun* are probably better known in the Muslim world than *Romeo and Juliet* in the Christian world. Meanwhile, the real control of the Islamic world east of Baghdad, where the Caliphs lived and from where they supposedly ruled, fell gradually in the hands of Persian-speaking regional rulers whose own local courts often employed Turkic soldiers and servants bought from the Asian steppes east of Bukhara and Samarkand. Most of these Turkic groups adopted the Persian language and culture; some of them also founded their own dynasties. Notable among them were the Gazny rulers who thrust into the Indus River region repeatedly around 1000 C.E. and, at its height with Nishapur as capital, controlled eastern Iran, Khurasan, Afghanistan, and the Indus Valley.

As more and more Turkic-speaking groups were drawn from the Asian steppes and as they converted to Islam, they also carried with them some features that were not in the traditional form of Islam. Many were won over by mystic Sufi preachers, who roamed the steppes and lived among the nomadic Turkic tribes who on the whole practiced Shamanism. These indigenous beliefs and customs in turn helped to shape or modify some of the Sufi rituals among the Turkic-speaking Muslims. Most significantly, the Turkic groups started to use their own language in official communications as well as in science and literature. The rise of Turkic power was reflected in the western push by a succession of Turkic tribal groups and the replacement of Persian rulers along the way. Thus, by the eleventh century, the Turkic groups ranged from the Seljuk Turks in Anatolia to the Karakhan Dynasty, whose power reached Kashgar in today's China. Turkish pride was fully expressed when the lexicographer Mahmud of Kashgar wrote in the preface of the first Turkish dictionary, "I have seen that God has caused the sun of empire to rise in the house of the Turks"

Enter the Mongols

As the Islamic world expanded continually in geography and in ethnic and linguistic diversity despite the ebbing of political power of the Abbasid Caliphs, who were successively controlled by warlords (who called themselves *emirs, maliks, sultans*), all the Arab, Persian, Turkic, Greek, and Slavic populations, plus the Crusaders from Western Europe (*Franks*) and the Mamluks (slave soldiers usually from outside of the Islamic domain), were soon to be shocked and awed by a nomadic Mongol tribe that arose from the region between Lake Baikal and the Altai Mountains.

In three separate expeditions spanning about forty years, Genghis Khan and his sons and grandsons conquered with lightening speed Khwarasm (in Transoxania), Khurasan, west Iran, Mesopotamia, Caucasus, Anatolia, the Balkans, and south Russia. The Mongols subjected the conquered peoples to terrible atrocities at first, but later settled in and ruled the conquered lands with considerable skills by enlisting the help of local elites. This proved a Chinese saying, "The Empire can be conquered on horseback, but it is impossible to run it from horseback."

A grandson of Genghis Khan, Hulagu, captured and sacked Baghdad in 1258. He executed the Caliph and had hundreds of thousands of its residents killed, thus brought the Abbasid Caliphate to the end. After that, Hulagu was named Il-Khan to rule the Persian lands between the Amu River and the Euphrates, along with the Khan of Chaghatay in

Transoxania and the Khan of Golden Horde in north Caucasus and south Russia. Initially these Mongol khans were Buddhists, but after one or two generations they all converted to Islam. They thus gradually became independent of the Great Khan in China, but continued to promote and protect trade with China. At the same time, some of them also became great patrons of art and craft which benefited greatly from exchanges between China and the Khanates. The Chinese influence on painting in the Persian and Turkic lands was particularly visible because of the nature of this art form.

In the late fourteenth century, a Turkish-speaking Mongol of humble origin named Timur (known in Europe as *Tamerlane*) repeated what his Mongol ancestors did – first terrorizing Central and West Asia, then ruled with considerable success, again with the help of local elites. The difference between Timur and the earlier Mongols was his total allegiance to Islam. The irony, however, was that Timur was quite ready to slaughter his co-religious populations who dared to oppose his military campaigns. He massacred tens of thousands of residents in Isphahan and had a tower made of their skulls, and that was almost a repeat of what Hulagu did in Baghdad. Again, with ironic similarity, the sons and grandsons of Timur became cultured rulers and patrons of the arts and science in their respective territories.

The discerning reader must have noticed my long digression from a personal account of my interest in Islamic culture to a review of history. I did this because the initial urge for me to write this article came from "*My Name Is Red,*" a story about miniature painters in the Ottoman court in the late sixteenth century. I shall revert to this book later, but I have to say first something about "*The Adventures of Hajji Baba of Ispahan,*" which accounts for the first part of the title of the present article.

A Persian Odyssey

Shortly after I got interested in Islamic culture and at a time when the West was still trying to recover from the shock of the Iranian revolution inspired by Ayatolla Khomeini, I bought a book at McGill University Library's book sale in 1983. It was James Morier's *The Adventures of Hajji Baba of Ispahan*, first published in London in 1824. The copy I bought was a Random House edition of 1937 "with a profusion of illustrations by Cyrus LeRoy Baldridge," who traveled in Iran for one year to prepare for the illustrations.

The son of an English merchant in the Ottoman port of Izmir, James Morier later served as a diplomat in Persia where he traveled widely between 1808 and 1815. The creation of Hajji Baba, a barber by trade whose fate took him all around the Persian land and whose fortunes underwent drastic ups and downs in this picaresque novel, brought Morier considerable fame and fortune at a time when there was a great deal of European interest in the Orient, but few Europeans knew much about it.

I devoured the book in as little time as possible, but it was still not a quick read, as I did not have the background information on the peoples, lands and social customs described in the humorously written book. After reading it, I thought I had learned quite a bit about the Muslim way of life in general and nineteenth-century Persian life in particular. I did detect a tone of superiority in the author's depiction of the conniving, corrupt and greedy nature of some of the characters, including Hajji Baba himself. Just like nineteenth-century European readers, I had no reference point with which to judge the truthfulness of what was written or illustrated.

I was, however, not without any sense of history. When I read the remarks made by Baldridge, the illustrator of *Hajji Baba*, who compared old Persia with twentieth-century Iran, I found his views dubious and disturbing. "A century has passed since Morier's time," says Baldridge; some of the old things in the Persia of Hajji Baba have been replaced by modern equipment. Then he laments the passing away of an exotic culture and says: "'Oriental' is synonymous with 'glamour.' Persia is an oriental land, long symbolized by flower strewn rugs, lyrical descriptions of Persian gardens, gay miniatures, and old tales of people whose garb was lovely and fantastic. In reality, the flowery fields of rug and miniature memorialize the wondrous – but brief – miracle of spring in a half empty land whose starkness often beats harshly upon the eye. And, in the name of modernism, the old costumes have been abolished by government decree." Even though at the time I had not heard of the now famous work *Orientalism* by Edward Said, I could not help but ask a question of my own: Why is it that the author Morier seemed to poke fun constantly at the "funny", i.e., backward behavior of the Persians, Kurds and Turkomen in his novel in early nineteenth century, but the twentieth-century illustrator Baldridge lamented the disappearance of old costumes and lovely and fantastic garbs?

Now, twenty-five years after the Iranian revolution, I wonder what the illustrator Baldridge would feel if only he could know what happened to what he saw as Iran's forced modernization by government's decree. Would he agree with the Shiite mullahs in Iran (or for that matter, the Sunni Talibans in Afghanistan) who proclaimed that they wanted neither the East nor the West, but only Islam? Would he be happy to see today more traditional costumes in Iran (and

Afghanistan) than when he traveled in Iran in the 1930s, and possibly when *Hajji Baba* made his Persian odyssey?

The Magnificent Ottomans

The Ottoman Turks, who succeeded the Seljuk Turks in Anatolia as a Sunni Muslim power around 1300 C.E., captured Constantinople and ended the Byzantine Empire in 1453 C.E. under Sultan Mehmet the Conqueror. At the zenith of the Ottoman power during the rule of Suleyman the Magnificent between 1520 and 1566, the empire's territories included southeastern Europe, the Middle East, and North Africa. Ottoman troops laid siege on Vienna twice (1529 and 1683) in the west, and, in the east, captured Baghdad (1535), controlled the whole of today's Iraq, even occupying Tabriz of Safavid Persia for some time. The Ottoman culture was also at its Golden Age during the sixteenth century and the first half of the seventeenth. Plane trees, tulips and sweets that are today associated mostly with France, Holland, and Austria respectively, were commonplaces in Istanbul during its splendid days.

The Safavid Empire of Persia, with its state-orthodoxy of Shiite Islam and long shared borders, was for a long time a major enemy of the Ottomans who considered themselves the protector of Sunni Islam and kept the Caliphs in Istanbul. Beyond the Safavid Empire to the east in Hindustan was the Mughal Empire whose rulers were descendants of Timur and Genghis Khan, but whose court language was Persian. Despite rivalry among them, these three Islamic empires shared a very similar cultural tradition.

To the west of the Ottomans, Europe in the late sixteenth

century was at the height of the Renaissance and going through a rapid phase of expansion in trade and territorial acquisitions. The Franks, as the Europeans were known in the Muslim world, were both competitors and trade partners of the Ottomans. They viewed the Ottoman Empire with both fear and contempt, and considered the Ottoman way of life exotic but backward. The Ottomans also had their complex feelings toward the infidel Franks whom they treated with suspicion, but also begrudging admiration. Many Ottoman nobles made visits to Europe and were impressed by what they saw, including paintings by Titian, Tintoretto and Veronese.

During the rule of Sultan Selim II (1566–74), peace treaties were signed with Austria and Persia. However, a series of wars occurred again between the Ottoman Empire and the Safavid Empire during the rule of Sultan Murat III (1574–95), who, besides leading the war efforts, was also a great patron of miniatures and books. It is against this backdrop and in the setting of the workshop of the miniaturists and bookmakers of Sultan Murat III that *My Name Is Red*, Orhan Pamuk's marvelous novel, unfolds its fascinating story.

To commemorate the one thousand years of Hejira (1591 C.E.) according to the Muslim calendar, and to make an impression on the Franks, the Sultan wanted to present the Doge of Venice with an illustrated book on the Sultan's own successes, including a portrait of himself drawn in the European style. Due to ideological differences and personal jealousy, one miniaturist was murdered, setting up a very intriguing and tense drama in and around the palace workshop. The author, however, very cleverly made the protagonists of this novel, Black and Shekure, not miniaturists, but two cousins whose love story had a faint and dreamy resemblance to the famous story of Khusraw (spelled Hüsrev in modern-day Turkish) and Shirin.

If I devoured *The Adventures of Hajji Baba of Ispahan* some twenty years ago without any knowledge of the Persian way of life, I savored *My Name Is Red* slowly in small pieces, often re-reading a passage several times to make sure I understood not only the intriguing storyline, but also the author's rendition of art theories as well as the undertone of his discourse.

In a delightful and masterful manner, Pamuk uses the voice of the first-person in all 59 chapters, seemingly in the style of a traditional storyteller. But the book most definitely does not tell a traditional story. It is historical romance, detective story, psychological thriller, art theory, and religious polemics all fused together under a storyteller's cloak, but with a postmodernist's spirit. The familiarity of the author with the art of miniature painting and the life in Istanbul in the sixteenth century can easily mesmerize the reader into an indulgence in the bygone glories of the Ottoman sultans, viziers, pashas, sufi masters (*hojas*), Janissary soldiers, and, of course, miniaturists. Yet the book is anything but nostalgic. In describing in detail the lives of the characters in the book and by making them go into self-analysis, the author deftly weaves a piece of the Ottoman social fabric and lifts the hypocritical facade of its political and religious structures. His good humored protest on behalf of dogs for their treatment in the Koran and the hadith is an example of his skillful needling of the Muslim tradition, an act that would easily incur the wrath of the conservative elements within Islam. Likewise, the frequent reference to homosexual behaviors and erotic acts would not go well with the pious Muslims. This, I am sure, Pamuk did this with a purpose in mind.

Unlike the nineteenth-century English author James

Morier, who sometimes ridiculed the Persian customs in an open and blunt manner, thus earning him a mention in Said's *Orientalism*, Orhan Pamuk, who has lived all but three years of his life in Turkey, is immune to accusations of having an "Orientalist" attitude and therefore able to expose from within the Orient the feebleness and wickedness of a medieval society, towards which, regrettably, many Muslims today seem determined to march. To those who proclaim "Islam is the solution," the depiction of life in the most magnificent days of the Ottoman Empire could serve as an antidote, whether this is indeed Pamuk's intention.

Seeing Is Believing

I believe the most distinguishing feature of *My Name Is Red* lies in its in-depth discussion of the art of miniature painting. If one suspects that Turgenev wrote *Fathers and Sons* in order to discuss nihilism, Orhan Pamuk probably wished to discuss the art of miniature painting when he conceived this novel. No less than 16 out of the 59 chapters in this book are devoted to either a description of the art of illustration or the history and theories of this art.

In the Islamic tradition, calligraphy was an approved and more respected form of art than painting because of the strict prohibition against idolatry. Painting is inherently dangerous if one starts to depict objects that can conceivably be worshiped. However, paintings from the Byzantines, the Hindus, and especially the Chinese simply could not be stemmed. Thus painting always existed in any Islamic society and was more or less accepted except by the most ardent religious groups.

To discuss painting it is essential to begin with the

material on which painting is executed. With paper spread from China to the Muslim world, painting and bookmaking were made much easier; therefore paper had a huge impact on the development of art in particular and of culture in general in the Islamic world. This is of course true of the situation in Europe as well. In 751 C.E., an Arab army based in Samarkand defeated an expedition force of the Tang Dynasty in Talas (near Tashkent) and captured many Chinese artisans, including some papermaking craftsmen. The Arabs subsequently established a factory in Samarkand to produce paper. Since that time, paper gradually replaced parchment as the preferred material for writing in the Islamic world. As the technique of papermaking later spread gradually westward, paper began to be made in Spain around 1150 and in France around 1200. Up until the time of Sultan Murat III of the Ottoman Empire, paper made in Samarkand still enjoyed a reputation for its special quality.

Next to the material for painting, the painter's tools are also of great importance. Brush painting was probably introduced from China to Persia very early on, but the Mongols played a key role in facilitating the diffusion of this art form in the Islamic world, particularly Persia. According to a story recounted by Pamuk in his novel, the famous calligrapher Ibn Shakir, by hiding on top of a minaret, saw what happened during the pillage and massacre by the Mongol soldiers when they overran Baghdad in 1258. He felt instantly an urge to record what he saw and started to depict the scene on the ground from his high vintage point. After he came down from the minaret, he abandoned his belief that calligraphy was the best form of art, and wished to learn the art of painting. So he began to walk eastward in the opposite direction of the Mongol troops in order to find a Chinese master.

The authenticity of this story of course cannot be established, but there is no doubt that Persian painting (including ceramic making which involves painting) from the thirteenth century onward until European painting was introduced in the sixteenth century showed unmistakable Chinese influence in terms of tools, colors, motifs, and composition. During the successive Timurid rulers, there was further development of Persian painting with distinct Chinese influence. In *My Name Is Red*, there are numerous references to the masters from Herat, Shiraz and Tabriz in the fifteenth and sixteenth centuries. Their migration from the eastern capitals to the western cities helped spread techniques of Chinese masters. Pamuk and other art historians cite the use of curling clouds, the way mountains and trees are drawn, the introduction of dragons and phoenixes in the paintings, the use of black ink and fine brushwork as examples of Chinese influence on Persian painting. The Ottomans and later the Mughals invited many Persian masters to their own courts and had them teach gifted local apprentices.

Visual art, from ancient cave drawings to digitally manipulated images today, represent the most vivid and credible form of communication. Its power of persuasion did not escape the attention of generations of religious and secular leaders who wished to use painting as a form of education or an instrument to propagate certain ideas. In the case of Islam, this form of education or propaganda was underdeveloped until the thirteenth century due to prohibition of idolatry.

Reading *My Name Is Red* and examining some Persian miniature paintings, I noticed two perplexing features. First, the Persian painters either totally adopted the Chinese standard of beauty in facial features or were still under the sway of the Islamic aversion to human portraiture. The faces in many Persian paintings were with high cheekbones and

Chinese eyes. I wish I could hear a more satisfactory explanation of this curious cross-cultural phenomenon than my conjectures here. Second, although the Islamic and Chinese senses of esthetics differ a great deal, the majority of both schools of painters did not attempt to depict directly what their eyes saw, but what their mind conceptualized as the appropriate forms of representation. Therefore, proportion in height and distance mattered less than the importance of the subject in the painting. To a Chinese artist, this type of rendition is what the Chinese literati called *yijing* 意境 (artistic conception), an expression of the artist's values and ideas rather than a reflection of the reality he perceives. For a devote Muslim painter, his work was not to reflect what he saw but what he believed Allah would like the viewer to see. In *My Name Is Red*, an old master miniaturist in Herat said to Ulüg Bey, grandson of Timur, "Allah created this earthly realm so that, above all, it might be seen . . . painting is the act of seeking out Allah's memories and seeing the world as He sees the world." With these divergent motivations, how coincidental it is that the Chinese and Persian painters could agree on the method of organization in a painting. As a consequence, there is little impetus to seek a new method to depict a scene or an object from the position of a given observer, or to use perspectivism developed by European painters during the Renaissance. (It should be noted, however, that perspective painting had been made by some Chinese artists before the eleventh century, but this was largely abandoned when the literati style of painting gained pre-eminence in China.)

Orhan Pamuk went into great length to argue that painting by an artist's hand is an act out of memory rather than drawing what is actually seen by him. A strong proof for him is the fact that the artist cannot at the same instant look at the subject of his painting *and* look at the paper on which his hand draws. That is why, so the theory goes, many Persian masters could continue to paint horses after they went blind, for their hands were so used to the strokes of drawing a horse that it was not necessary for them to see a horse. This assertion may require neurophysiologists, cognitive scientists, cultural anthropologists and social psychologists to validate, but I do know there are blind or blind-folded calligraphers in China who can do excellent calligraphy without looking.

In *My Name Is Red*, there is also a great deal of discussion of the true meaning of the "style" of an artist by which his work becomes recognizable. To this question, Pamuk provides an amazingly simple but almost nihilist answer: the "style" of an artist represents nothing more than his "imperfection." The slit nostril of a horse, an important clue in the murder case at the beginning of this novel, is an unintended imperfection of an artist or his subconscious "signature." Along the lines of this thesis, I would say that this very idea might be Pamuk's tongue-in-cheek "signature" in his successful book, which, notwithstanding his intentions and despite his vast and detailed knowledge on the subject, should not be read as a treatise on the styles of painters. Yet, I do not think this is an imperfection on his part. It is just one more proof of the ease with which he can find a good mélange for his imagination, his observations, his sense of humor and his serious discourse on art and society to produce an entertaining and yet solidly meaningful piece of literature, as easily appreciated as the art of the master miniaturists he portrayed.

Pamuk refers to a romantic legend repeatedly in *My Name Is Red*, i.e., upon seeing a picture of the handsome Persian prince Hüsrev (Khusraw), the beautiful Armenian princess Shirin immediately fell in love with him. This legend illustrates first of all the power of visual representation and, secondly, it seems to argue for semblance or similitude in portrait painting. With the march of time, the resistance to realistic representation notwithstanding, Chinese and nearly all Muslim artists have adopted the European style of portraiture, a style that engenders a political plot in *My Name Is Red*.

More important, this romantic story tells me that there is an inexplicable connection between the eye and the heart. We mortals are all, knowingly or unknowingly, mere implements of our heart. For me, the sight of the *Alhambra* in Granada immediately attracted me to the Islamic art and the histories of Islamic societies. My eye contact with a copy of *My Name Is Red* last summer in Hong Kong has given me not only enormous pleasure in reading it, but also the urge to meet its author in the cold winter of Istanbul. And that meeting with this accomplished Turkish author is also why I now have the courage to recount, timidly and amateurishly, my personal journey in the last twenty years into a fascinating yet extremely complex realm.

Note:

1. Both the Persian and the Tukic languages contain many words and proper nouns of Arabic origin; however, due to different pronunciations in these languages, they appear to have different spellings when transliterated into the Latin alphabet. Moreover, there exist multiple transliteration systems for Arabic (as well as the various dialects of the Persian and Turkish languages), making it difficult for an author in English to choose a single system of transliteration and for a reader to know the differences. For example, the scripture of Islam can be spelled in English as *Qu'ran, Quran, Kuran,* and *Koran*. In this article, I have tried to use what I think is the most common spelling of names and places with Arabic, Persian, Turkic and Mongol origins; no special efforts are made to

adhere to a given system of transliteration. However, when direct quotations or proper nouns in current usage are involved, I use the exact spelling of the source. For example, I use *Koran* simply because it is the more customary spelling in English. For the book by James Morier, I use the form *Hajji Baba,* because that is what he used, but for the restaurant in Istanbul I use *Hajj Baba* because that is what the restaurant calls itself. For these reasons, the city in Iran is variably spelled *Isfahan, Isphahan* or *Ispahan.*

(October 2004)

6

"The Accidental Tourist" in Egypt

I have been in Egypt for one month now, longer than most tourists spend here, on an academic visit to Cairo University, the most prestigious university in Egypt. They call it the "mother university." It was founded in 1908 by Egyptians during the days of British rule and it has twenty-three faculties with over 190,000 regular students and a total teaching staff of more than 7,800.

Professor Ali Abdel Rahman, President of Cairo University, prodded by Madame Somaya Saad, the Egyptian Consul-General in Hong Kong, issued the letter of invitation to enable me to come here for one month and enjoy the use of a spacious office suite under the friendly gaze of President Hosni Mubarak from the wall behind me. It is now seven o'clock in the evening on Sunday (Friday is the day off here) and the office building is emptying out. But the streets below are still choked by cars and teeming with pedestrians. Not in any hurry to have dinner yet because I had a full lunch at three o'clock, as is the custom here, I find myself in a reflective mood.

More than five thousand years ago, a brilliant civilization sprang up along the Nile. Anyone who has visited the pyramids and the ancient temples cannot escape the question, "Why were the ancient Egyptians so advanced?" When one compares the life of Egyptian farmers and laborers today, many of whom still live in mud shacks and use the donkey for transport, to the way their ancestors lived, one is naturally led to the next question, "Is this country endowed or burdened with so much history?"

The brilliant ancient Pharaonic civilization was successively conquered by the Greeks (Alexander and Ptolemy) and Romans (Julius Caesar and Otavia); from the third century onward Egyptians (Copts) became Christians. But Egypt was again invaded by the Arabs in the seventh century, thereby starting the dual process of Islamization and Arabization. Cairo was built about a thousand years ago and its Arabic name, *Al-Qahirah,* means City of the Victorious, because the Fatimids, a Shiite group who came out of Libya and Tunisia, seized Egypt from the Abbasid Caliph based in Baghdad. Since the Mongols exterminated the Abbasid Caliphate in the thirteenth century, Cairo, under the successive rule of the Mamluks, Ottomans, French, British and the Egyptian republic, has been the center of Arab culture. The Al-Azhar Mosque in Cairo has been the standard-bearer of Sunni Islam and fountain head of different schools of thought. It was at Al-Azhar that the celebrated fourteenth century historian Ibn Khaldun proposed his theory of the cyclic development of cultures, a theory which gave inspiration to the twentieth century British historian Arnold Toynbee who argued that the twenty-first century would belong to the Asians.

When Napoleon invaded Egypt in 1798, Egypt, along with the entire Muslim world, was traumatized, but it also became inseparably entangled with an aggressive and dominating Western Europe. Under Mohamed Ali, an Albanian who ruled Egypt in the name of the Ottoman Sultan, but in reality on his own, Egypt began its drive for modernization in the first part of the nineteenth century, earlier than the modernization movements of Japan and China. It sent many students to Europe and established a

"European-style" army as well as a national post system. The many magnificent, but now aging, edifices in Cairo are testimonials of this epoch to Western-oriented development, which was financed first by cotton exports and later by Suez Canal revenue.

Religious tolerance and diversity have long been a tradition in Egypt. Many Coptic churches and a few synagogues stand in different parts of Cairo today. Copt Christians, inhabitants of Egypt before the Arab conquest, now constitute about 5 percent to 10 percent of the population. The former Secretary-General of the United Nations, Boutros Boutros-Ghali, is a Copt Christian and his father was once Prime Minister of Egypt.

And history continues to unfold. Since I arrived in Cairo, several events of considerable significance have taken place.

At the urge of Western countries, the ruling party in Egypt, led by Mubarak, liberalized the parliament election this year, and opposition parties were free to run campaigns. After three rounds of voting, the banned but tolerated Muslim Brotherhood took roughly 20 percent of seats, more than anyone had predicted and a four-fold increase from the last Parliament. While the government party managed to maintain its two-thirds majority, the secular opposition parties all lost badly. Women and Christians fared even worse. Members of the Muslim Brotherhood are not extremists, but they are avowed Islamists and their most famous slogan has been "Islam is the solution." In an atmosphere of high unemployment and general frustration felt by ordinary Egyptians, the Islamists have now gained a significant platform and will have a much larger audience in the future.

Strolling around campus, I have taken my own "head count." About three-quarters of all female students wear headscarves, a sign of Muslim pride in recent years, suggesting that interest and fervor in religion are on the rise among the well-educated young people here. Since Egypt is, by any measure, the center of the Arab world, what happens here will have far-reaching consequences in all Arab countries.

Whatever the level of religious devotion, Egypt has over the years accumulated a sizable pool of human capital. Besides a large corps of administrators, military officers, doctors, engineers, lawyers, bankers, accountants and thousands of tourist guides fluent in all kinds of languages, including Chinese, Egypt has also produced four Nobel Laureates. The first, Mohamed Anwar El-Sadat, President of Egypt (1970–1981), received the Peace Prize in 1978. He was a Military College graduate. Naguib Mahfouz, who won the Literature Prize in 1988, was educated at Cairo University. Writing about the life of ordinary Egyptians with an anti-tradition stance, Mahfouz celebrated his ninety-fourth birthday a few weeks ago. The third Egyptian Nobel Laureate, Ahmed Zewail, won the Chemistry Prize in 1999. He graduated from Alexandria University in Egypt and got his doctorate degree from the University of Pennsylvania, U.S. The fourth, Mohamed El Baradei, Director-General of the International Atomic Energy Agency, won the Nobel Peace Prize in 2005. He is also a Cairo University graduate.

Also at about the same time, Egyptian surgeons performed a lung transplant involving a live donor. These all speak well of the caliber of the Egyptian higher education system with which people in Hong Kong are largely unfamiliar. In this sense, I am glad that City University will receive some students from Cairo University next year and that important academic collaborations are underway.

Egypt is an enchanting and yet perplexing country. While the Parliament elections and the Nobel Peace Prize are of historical significance, ordinary Egyptian seems to care much more about the football match between Egypt and Saudi Arabia held in Tokyo in early December. As I listened to the thunderous cheers and loud sighs, from street corners to shopping malls, I sensed that the ancient land and modern state of Egypt mean a great deal to its citizens.

The impact of modernity has produced varied responses from the Egyptian society. The twentieth century saw Egyptian nationalism, Pan-Arabism, and Islamism compete for the hearts and minds of the Egyptian people. Nearly all citizens of Egypt are culturally Arab and over 90 percent of them Muslim. Of the three identities they share, i.e., Egyptian, Arab and Muslim, which one will be most convincing in shaping their destiny? The answer to this question is still unclear at this moment but it may foretell the shape of the Middle East tomorrow.

(December 2005)

7

From Zhang Qian's Western Expedition to the "Factory of the World"

The Formation and Development of Civilizations

Human beings entered the "Age of Civilization" about ten thousand years ago, when the two productive activities of planting and husbandry began to be practiced in different regions of the world. From that time on, some communities also engaged in intellectual activities such as writing and mathematics. These are the common characteristics of all of the world's civilizations.

The French historian F. Braudel believes that geography has had a deep influence on the development of civilizations. For that reason, each civilization can be identified by a place on the map. The distinction between agricultural and nomadic civilizations lies principally in their natural environments. Because the geographic and human environments do not change easily, it is impossible to observe through a limited time frame the pattern of development of a culture that takes a long period (*longue durée*) to evolve.

The British historian A. Toynbee compares the rise and decline of various civilizations in history and concludes that

---

This article is based on a speech that the writer delivered at the International Symposium on "East Asia in the Twenty-first Century: Cultural Construction and Cultural Exchange" at Peking University on April 19, 2006.

the main factor that determines the development of a civilization (or a society) lies in the way it responds to the challenges of the external environment, be they natural or man-made.

Both Braudel and Toynbee believe that interchange and mutual borrowing exist by necessity between civilizations, but each civilization has its peculiar inclinations. Further, the facility of communication between civilizations determines the speed and depth of cultural exchange. For example, paper-making technology from China reached Central Asia by the mid-eighth century, and after that, made its way through Bagdad, Cairo and other places, but it was not until the thirteenth century that paper mills appeared in France.

Zhang Qian's Journey to the Western Regions

East Asian civilization, represented by Chinese culture, came into being at the eastern end of the European-Asian landmass. Blocked by the Altaic Mountains, the Pamirs and the Himalayans, it was isolated from other ancient civilizations, and for that reason, retains some very distinctive characteristics, such as the writing system.

The earliest record of communication between Central China and the so-called Western Regions (*Xiyu*) is to be found in Zhang Qian's attempt at "opening holes through the blockade" (*zao kong*) in the second century B.C.E. This was a conscious effort on the part of the Han Empire to seek contact with the world outside in response to the threat posed by the people in the northern steppes (*Xiongnu*). After Zhang Qian's expeditions, formal contact between Chinese and Central and

A Tang dynasty painting showing the Emperor Wu's blessing on Zhang Qian's expedition to the Western Regions.

Southern Asian civilizations was established. What Emperor Wu of the Han did not foresee at the time was that Buddhism, and all of the artistic forms that came with it, would thereby be brought to China, transforming the spiritual world and artistic life of the Chinese people, and in turn affecting the development of the entire East Asian civilization.

In addition to continuous east-west exchange, there existed also frequent contacts between north and south in the European-Asian landmass from the second to the sixth century C.E, manifested in repeated invasions by nomadic groups of the sedentary agricultural societies in Europe and West, Central and East Asia.

In Europe, these invasions brought complete ruin to the urban civilization and social order established by the Roman Empire. In West and Central Asia, Persian civilization withstood pressure from the Turkic groups and, moreover, steered them to agricultural and commercial activities. In East

Asia, the high population density and advanced civilization of the Han people made it possible to absorb and assimilate almost all the invaders who came from the steppes over several centuries, bringing about a melding of peoples unprecedented in history.

The Chinese culture of the seventh century C.E. incorporated cultural elements of Central and South Asia with the results of the melding of the various groups from the steppes. In the several centuries that followed, China displayed creativity hitherto unseen in areas such as technology, literature, philosophy and the arts, and put in place social and economic innovations. The new energy evident in poetry, calligraphy and painting, the practices of the civil examination system and bank draft, and the invention of printing are examples well known to all.

Europe Rules the World

Between the eighth and thirteenth centuries C.E., the Islamic region wedged between East Asia and Europe provided a land passage that connected the two; Arabs and Persians also came to China by sea for commerce. On the whole, Muslim merchants can be said to have served as cultural mediators between East Asia and Europe in medieval times. During the thirteenth century C.E., Mongols invaded West Asia and Eastern Europe. Chinese civilization became all the more familiar to people in Europe as a result, and interchange between the two ends of the European-Asian landmass became all the more frequent.

After the "dark age" that lasted for almost a thousand years, there gradually emerged in Europe a social order completely different from that of the Roman Empire. At the same time, partly as a result of the challenge of the Mongols and the Islamic forces (especially the Ottomans), and partly as a result of the incorporation of elements of cultural sustenance from the Islamic world (especially Spain), various places in Western Europe underwent the Renaissance, the discovery of navigation routes (and the conquest of overseas territories), the introduction of the capitalistic mode of production, and religious reform movements from the fourteenth to sixteenth centuries C.E. These are early signs of the rise of Europe. As early as the seventeenth century (i.e., the Newtonian age), Europe began to demonstrate characteristics of what many scholars call "modernity." By the nineteenth century, it had come to dominate the entire world.

That European civilization could dominate the world in the nineteenth century was primarily due to the results of the Industrial Revolution in the eighteenth century. All ancient civilizations which have felt the impact of European industrial civilization have willingly copied the production technology of industrial civilization and at least some of the accompanying systems, such as the establishment of modern schools and modern armies.

Even though Muslim societies in North Africa, West and Central Asia also imitated to a certain degree the dominant culture of Europe, their fundamental response was to reaffirm their belief in Islam. This of course is a self-defense mechanism that comes from a reluctance to be subjugated by the Other. On the other hand, the Islamic cultural tradition is indeed the primary force that is capable of motivating entire societies to work towards modernity.

South Asia and Southeast Asia gradually became colonies of the European powers. People living in these two regions in the nineteenth century, be they Hindus, Muslims or Buddhists, accepted European science, technology and many other features of social organization, but members of these ethnic and religious groups also maintained a kind of cultural awareness, with the hope of advancing towards modernity while preserving the characteristics of their own culture. Such attempts have not yet brought about any remarkable results, as the effort made by people in the core Islamic region can clearly demonstrate. As a matter of fact, industrialization is far from being common in these regions even to this day.

Because East Asia is the last region that came under the impact of European forces, it has had more time to observe and critique Western civilization. Even though the specifics may vary from place to place, by the beginning of the twentieth century, people in China, Japan and Korea had shown a willingness to accept industrial civilization and, to varying degrees, the European social system (such as the parliamentary and legal system) and aspects of life and customs (such as attire and social etiquette). Nevertheless, the resilience and pre-existing value system of East Asian civilization also made themselves felt in two ways: first, the acceptance rate of Christianity among ordinary people in East Asian countries is still relatively low; and second, the elites of various East Asian countries, including Japan, the first in the region to adopt parliamentary democracy, still tend towards internal consensus at the upper levels rather than open confrontation.

If one compares the Islamic culture of West Asia, the Hindu culture of South Asia, and the Buddhist and Confucian culture of East Asia, one will see an interesting phenomenon: Islamic culture, which has the deepest historical relationship with Christianity, has the longest period of contact with Western Christian culture, and which is closest to Europe geographically, is most resistant to Western civilization. On the contrary, East Asian culture, which shares no historical relationship with Christianity, has the shortest period of contact with Europe, and is geographically most remote from Europe, has been most ready to accept Western culture.

East Asia: the "Factory of the World"

Today, various economies of East Asia – those of Japan, Korea, Taiwan, Hong Kong, and the Chinese Mainland – have all reached a certain level of industrialization. The quality of their products is high, the quantity considerable, and the cost lower than in Europe and North America. In the 1980s, Japan was called the "Factory of the World," a name that has now been transferred to the Chinese Mainland. In fact, it would be very appropriate to use the same name to describe the "Four Little Dragons in Asia" in the 1990s. At present, electronic products that require advanced technology, micro-computers, and telecommunication equipment (such as mobile phones) are mostly produced in East Asia. The manufacture of traditional large-scale means of transportation, such as ocean-going vessels and automobiles, has slowly been taken over by Japan, Korea and China. The first six of the ten biggest container ports in the world today are Hong Kong, Singapore, Shanghai, Shenzhen, Pusan and Kaohsiung. All this goes to show that East Asian culture has responded to the challenges posed by European industrial civilization in a more effective way than other civilizations.

A. Toynbee predicted fifty years ago that, because of the continuous "diffusion of technology" that took place from the nineteenth to the twentieth century, East Asia would become the most productive and technologically and culturally advanced region in the twenty-first century. Even though there are already signs that bear out Toynbee's prediction, the emergence of the "knowledge economy" and the worldwide division of economic activities makes it necessary for East Asian countries to respond to extremely crucial challenges before they can truly become leaders of the global age.

Creativity-based Economy

Until now, the most outstanding accomplishment of East Asian civilizations and societies in the process of modernization lies in their rise in productivity, which is quite different in nature from the flowering of Chinese civilization from the seventh to the eleventh centuries. Apart from the vitality it demonstrated in economic activities, East Asia (including Japan, the first in the region to industrialize) in the twentieth century witnessed hardly any major scientific and technological inventions or social or economic innovations. For half a century, the expansion of East Asian societies has been built upon the hard work of a well-trained and disciplined labor force. To be sure, private savings from the people and appropriate governmental actions, such as constructing and maintaining an excellent infrastructure, have also been very important.

Some scholars have pointed out that the early period of modern economic development was built upon investment, where governmental policies, the supply of capital and the quality of the labor force have all had an important role to play. The value system of East Asian civilization happens to be conducive to producing a well-trained and disciplined labor force. However, in the current phase of the global division of economic activities, the next level of economic development will be built upon inventions and innovations, with a larger profit coming from creativity. In this particular phase, societies (or civilizations) that succeed in promoting originality and cultivating a large number of creative talents will be at an advantage; to maintain this, one would require a social and cultural ethos quite different from that which produces a well-trained and disciplined labor force.

Without creativity as the new leading force, East Asian societies have earned the name "Factory of the World" for the last fifty years primarily on the strength of investment, and they have not yet moved to the forefront of the world as Toynbee so optimistically predicted half a century ago. Absent a social and cultural atmosphere that allows, among other things, the free flow of information to produce creative talents, it would be very difficult to arrive at a creativity-based economy. That is why a society that produces new talents and promotes creativity will become the "headquarters" of the global division of labor, and the region (or countries) capable of processing industrial and commercial data in English, now commonly used internationally, and providing superior and low-priced services will become the "Office of the World." East Asian countries, which have already shown themselves capable of manufacturing with high efficiency all kinds of products, will probably continue to play the role of the "Factory of the World."

The New Challenges
Facing East Asian Civilization

A civilization resembles a person. It can transform itself by learning, but it is also subject to inertia, and is not at all easy to transform.

The cultural traditions of all East Asian countries in the

last one thousand years have always emphasized obedience to elders, the superior, and the collective. Parents require their children to be "submissive," teachers want their students to be "docile," and governments show a definite dislike for the "unruly masses."

Creativity and obedience are two distinct states of mind. Creativity requires motivation from within and other qualities that come from a long period of cultivation. A society that wants to promote obedience and creativity at the same time will be posing a major challenge to itself. Singapore, culturally a part of East Asian civilization, has run into precisely this kind of difficulty in recent years: it is trying to promote creativity in its people through the authority of the government. Even though it is still unclear what the result of this attempt will be, one possible outcome is that, with full intention of obeying the government's bidding to be creative, its people will nonetheless fail to do so. The other possible outcome is that, while the people may become creative, they may not be so obedient to the government any more.

As I argued above, a civilization displays strong continuity. East Asian civilization has always stressed social order. The education of young people today, for example, still relies mainly on indoctrination and memorization. Cultural patterns such as this have taken a long period of time to form, and will not be changed radically through government policies in a short time. To develop creativity in the near future, to make their way to the "headquarters" of the new global economic order, and to shed the dubious honor of "Factory of the World" that depletes resources and damages the environment, East Asian countries have yet to construct an appropriate social culture, and be ready to put in some long-term effort.

8

The World at the Time of Zheng He

Introduction

In the early part of the fifteenth century (1405–1433), the Chinese explorer Zheng He embarked on a series of seven voyages West, taking with him a large fleet of ships and a multitude of officers and men (Fig. 1). Whether judged in terms of the time involved, the scope of territories reached, or the amounts of men mobilized for the task, Zheng's voyages are clearly an event of key importance, both in the history of China and of the world. And yet, due to the subsequent ban on maritime trade with foreign countries imposed by the Chinese court, Zheng's achievements were soon to be forgotten; his heroic undertaking seemed to have had little serious impact on the world stage. When one leafs through world histories published in a variety of different countries, each records the seafaring achievements of Columbus, Vasco da Gama and Magellan, yet few mention Zheng He's voyages or the impact of his missions to the Southern and Western Oceans.

When judged objectively from the perspective of the different cultures of the world and the development of society in different regions, it must be admitted that Zheng He's seven voyages did not occupy a dominant or even mainstream position in the historical development of China itself, since China at that time did not regard foreign trade as important,

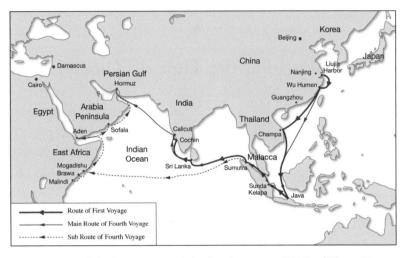

Fig. 1: Routes of the first (1405) and the fourth voyages (1413) of Zheng He.

nor did it need to occupy foreign territories for purposes of emigration or plundering resources. Moreover, since Zheng He pursued the Ming dynasty's policy of peaceful diplomacy, the places his fleet reached did not change the course of their subsequent social, economic and cultural development because of any armed attack. In other words, whilst Zheng's voyages were an unprecedented achievement, to the world as a whole, they seem rather a low-key affair. They did not have the same kind of earth-shattering impact as Genghis Khan's earlier western conquests or Columbus's subsequent discovery of America. Nevertheless, this in no way means that people – and Chinese people in particular – should not remember Zheng He as a pioneering navigator and outstanding diplomat.

A hundred years ago, when the powerful maritime nations had carved up almost the entire world and China was fighting for its very survival, Liang Qichao and other scholars of the day revived and commemorated the memory of Zheng He's achievements. This was an act entirely appropriate in its historical context, for it served as a rallying cry to the Chinese people. Today, as we celebrate the six hundredth anniversary of Zheng He's voyages, China is entering a period of peaceful development and rich opportunity, and the ocean is of crucial importance to this development in the future. It thus seems particularly appropriate to commemorate Zheng He at this time.

In order to gain a clearer and more developed view of the world at the time of Zheng He's voyages to the West – a world that was still not fully understood by people of the time – the present paper seeks to provide an overview of key features or events of different regions in the world of Zheng's time, beginning with an account of China and East Asia and tracing west through to Europe. It will not discuss regions such as America, Oceania or sub-Saharan Africa, which had had little or no contact with the Eurasian continent.

China

China in the early years of the Ming dynasty was unquestionably the world's foremost power, whether considered in terms of the technology used in agricultural and industrial production, the flourishing of trade, the standard of living among both the urban and rural populace, the refinement of culture, or military power. The number of officers and men and their family members, the abundance of materials supplied, and the management and communications of the fleet led by Zheng He all provide ample evidence of the greatness of China's combined national strength. Scholars have shown through textual examination that the total amount of silver expended on Zheng He's seven voyages amounted to less than five percent of the palace coffers. Such an observation demonstrates that whilst the expenditure on Zheng's voyages was indeed huge, when considered from the perspective of the entire Chinese economy, there were sufficient funds to support these seafaring and diplomatic expeditions.

Why did the Ming Emperor Yongle decide to have Zheng He make seven voyages, accompanied by so many ships and men? There have been various conjectures made through the ages, but no-one has been able to come up with any definitive answer. The Yongle Emperor had originally been Prince of Yan, responsible for defense against the Mongols, and he must certainly have had an understanding of the power of the Mongols and the Timurid empire of Central Asia. Zheng He's successive voyages to the Indian Ocean along known sea-routes already established in the Song and Yuan periods, and his visit to Persia, a country under the control of the Timurid empire, must also have involved strategic geopolitical considerations: either the plan would be to unite with the Timurid empire and contain the Tartar and Oirat Mongols on the flank, or to find some other power in the Western Ocean through which to contain the Timurid empire, or both. These conjectures are by no means original observations, and yet it would seem that they have not been seriously discussed before. Of course, clearing the way for maritime transport routes, encouraging tribute trade and advancing peaceful diplomacy were all reasons for Zheng He's voyages. Yet the geopolitical considerations should perhaps not be underestimated. (Fig. 2).

The "Kangnido Integrated Historical Map of Countries and Cities", made in 1402 (Fig. 3), already marks Hormuz, situated at the entrance to the Persian Gulf. This was the final point of Marco Polo's journey to escort a Yuan dynasty Princess from China to marry into the Il-Khanate, and it is not surprising that it is shown on this map. What is surprising,

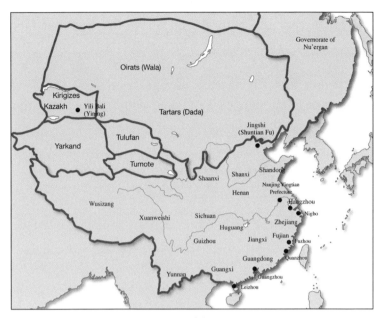

Fig. 2: Territory of the Ming Dynasty.

Fig. 3: Kangnido Integrated Historical Map of Countries and Cities (1402).

however, is that the map should show the Cape of Good Hope in South Africa, a fact which scholars have no way of explaining. Whatever the reality, this map at least demonstrates that Zheng He's voyages west were in no way like those of Columbus or da Gama, which aimed at finding new and previously undiscovered navigation routes.

East Asia and South-East Asia

Foreign policy in the early Ming dynasty chiefly consisted in getting rid of Yuan dynasty influence and gaining the acknowledgement of foreign countries. The most serious threats to Ming power at the time came from the Mongol power-base in the north and from Japanese pirates along the coastal regions of the south. Emperor Taizu made the north the key point for defense, and being unable to simultaneously wage war on the Japanese pirates in the south, he therefore made every effort to achieve a situation of peaceful coexistence with various overseas countries, as a means to forge a relatively settled international environment. Emperor Taizu once instructed a provincial official: "Now among the foreign barbarian countries, those that cause trouble to China must be punished, whilst those that do not cause trouble should not be attacked without provocation . . . I believe that those minor barbarian countries that are separated off by mountains or seas, and which are isolated away in a corner, do not constitute any source of trouble to China, and I have decided not to attack them." He lists as "barbarians not to be attacked" various countries including Korea, Japan, Ryuku, Annam, Zhenla (Cambodia), Thailand, Champa, Sumatra, Java, Srivijaya and Bo-Ni (Brunei); and he adds an "ancestral instruction" with the aim of preventing a situation in which "descendants in future generations, on the basis of China's wealth and strength, may thirst for battle victory, raising up arms when there is no enemy, and bringing about the loss of life."

In the East Asian milieu, Korea, Japan and Annam had all been profoundly influenced by Chinese culture. At the beginning of the fifteenth Century, the Korean peninsular was ruled by the Yi dynasty, existing as a vassal state of China, honouring the calendar of the Ming dynasty and paying tribute regularly. The Yi dynasty laid emphasis on Confucian learning, and in 1420 set up an academy for the pursuit of such scholarship. However, in order to preserve the characteristics of their own language, in 1443 they created the Korean alphabet, or "Hangul", the script that is still used today. Since the upper echelons of Korean society still placed importance on Chinese script, this set of alphabetic symbols was only to gain widespread use in the twentieth century.

Japan, by contrast, was an island, and thus whilst it was influenced by Chinese culture, it was not a vassal state. Kublai Khan's two attempts at conquering Japan in the Yuan dynasty had both failed, and this had caused Japan to become still more distanced from China. The beginning of the fifteenth century was the era of the Muromachi Bakufu, and the shogun Ashikaga Yoshimitsu (the grandson of Ashikaga Takauji) put an end to the north-south divide, building temples and palaces in Kyoto, including the famous Temple of the Golden Pavilion, or Kinkakuji.

Japanese trade with the various coastal regions of China had hitherto been characterized by a situation in which commerce and piracy existed side by side. In the early Ming, a policy of trade based on a tally-system was rigorously pursued, in which only those in possession of a tally issued by the Chinese government were entitled to come to China to engage in trade. This was to have an effect in regard to regulating tribute trade and eliminating the incursions of the Japanese pirates.

Annam (present-day Vietnam, then also known as Jiaozhi) broke free of China and established itself as an independent state during the Five Dynasties period in China, and the following Song and Yuan dynasties had not brought it back within their borders. In the Ming dynasty, at the time of Emperor Yongle, the Trần dynasty in Annam was overthrown and a request was made for China to send help in the form of troops. In 1406, Emperor Yongle duly sent troops to Annam, setting up a Jiaozhi Administration Commission. However, in 1418, another leader, Lê Lợi, staged an uprising, which led in 1428 to the establishment of the Lê dynasty, from which time Vietnam broke free of China and became independent.

Among the South-East Asian countries, some were still fishing and agricultural societies such as the Philippines, while others were beginning to become Islamic, and there appeared regimes governed by Sultans, such as Sumatra. After Zheng He's voyages, a total of thirty different countries' tribute ambassadors came to China, while eleven monarchs from four countries accompanied the tribute ships. Among these, the king of East Sulu died of illness in Dezhou in Shandong, on his way back to his country. The court gave him the funeral of a Ming prince and constructed a mausoleum which remains to this day.

At the southernmost edge of the Malay Peninsula, Malacca received an imperial honor in 1408, following which it broke free of Siamese control, its power gradually growing greater, to become a major commercial city. The Malayan language spoken in Malacca became a *lingua franca* for trade between the islands in the Southern Ocean, and was the origin of the official languages of present-day Malaysia and Indonesia.

In Sumatra and Java at the time of Zheng He there appeared a number of small Islamicized port states. Arab, Persian, Indian, European and Chinese merchants and pilgrims engaged in trade in the coastal regions, and economic activity was frequent. Aceh in Sumatra was the earliest stronghold of Islam.

In Siam, at the beginning of the Ayutthaya period, worship of Devarajas was introduced. Burma, Siam's western neighbor, during the era of the Mon kingdom, became one of the centers of Theraveda Buddhism, with its capital at Bago.

Throughout the Southern Ocean region, Chinese people settled. Some of these actually established military forces, seizing areas for themselves and plundering the wealth of passing merchants. One rather well-known example of such a ringleader is Chen Zuyi, who based himself in the Sumatran port of Palembang. Zheng He captured and sent Chen Zuyi back to China, while he pursued a policy of offering amnesties to others in return for their submission. In short, all the various regions in South-East Asia already had Chinese communities before Zheng He's arrival, which were made up of people from China's South-Eastern coastal provinces. When Zheng He went on his voyages, he gave them assurances and canvassed their views. This had a motivating effect on Chinese emigration to South-East Asia.

South Asia

Situated in the Indian Ocean, Ceylon was in ancient times known as the Country of the Lion Mountain. The south was controlled by a Buddhist kingdom established by two

Singhalese clans, whilst the north was controlled by a Tamil kingdom which believed in Hinduism. Jaffna in the north was a center of Hindu culture, with landlords controlling economic and political power. It is worth mentioning that the only instance of an attack giving rise to large-scale armed conflict during Zheng He's voyages occurred in Ceylon.

In the Indian subcontinent, the north, east and central areas were controlled by Islamic regimes (such as the Delhi Sultanate) that originated in Afghanistan. The southern area of the subcontinent saw the emergence of a stable and flourishing Hindu kingdom, Vijayanagar. Hinduism was created by the Aryan peoples who invaded the Indian peninsula in around the fifteenth century B.C.; they practiced the caste system as a means to govern people of different races and cultures. By the fifteenth century, the original Dravidian inhabitants of the south had long since been permeated with the Brahman culture of Hinduism. The kingdom of Vijayanagar thus brought together the Indian races with the new Hinduism, encouraging the use of Sanskrit, and leading to a period of flourishing and stability.

In the third year of Yongle (1405), the Ming emperor Chengzu decreed that an honorary title be conferred on the king of the Hindu state of Calicut, in the south of the subcontinent. A commemorative stele was erected by Zheng He, the text of which read: "We have traveled over one-hundred thousand *li* from China; the people here are settled and at peace, with customs not unlike our own. We carve a stone at this spot, to shine eternally for ten thousand generations."

Central and West Asia

After the Mongol conquests in the thirteenth century, the

whole of Central Asia, and Persia, Iraq and the Eastern part of Anatolia (modern Turkey) in West Asia, were respectively incorporated into the territory of the Chagatai Khanate and the Il-Khanate. By the middle of the fourteenth century, these Mongol rulers had already become united with the local inhabitants and had converted to Islam. However, in Central Asia, the renown of Ghengis Khan's "Golden Horde" and its descendants still persisted.

It was against this background that a Turkic general called Timur, styling himself a descendant of Genghis Khan, started out from his home base in Samarkand to conquer the entire territory of the Chagatai Khanate, invading East into India and sacking Delhi, then going further West to attack central Anatolia and the Caucasus. His conquest against the Ottoman Turks, who shared the same religion and ethnicity, may objectively be said to have postponed the fall of the Eastern Orthodox Byzantine Empire. In his latter years, in accordance with nomad custom, he divided the territory he had conquered amongst his various sons, whilst he himself returned to Samarkand and attacked eastwards into the Eastern Chagatai Khanate, situated in present-day Xinjiang. In 1405, Timur died on his way East to invade China. After his death, he was buried in the mausoleum which he had constructed for himself.

The huge empire bequeathed by Timur in the fifteenth century was the second most powerful empire in the world, after China. His son, Shahrukh, moved the capital to Herat, in present-day Afghanistan, building it into not only the political center of the Timurid empire, but also into the Perso-Islamic cultural center of the fifteenth century. Shahrukh's son, Ulugh Beg, was a ruler who emphasized culture and science over military and political affairs. Whilst he was Crown Prince, he also ruled Samarkand for a number of years, constructing an astronomical observatory there which was at the time the most advanced and precise in the world. He also built mosques, established religious schools, and turned Samarkand into a cultural center in which both Turkish and Persian were used.

Between 1416 and 1419, Zheng He reached Hormuz, at the entry to the Persian Gulf, which was at that time under the control of the Timurid empire. Did Zheng He liaise with special envoys from the Timurid empire, as touched on above? We shall not venture to provide any definitive answer. According to historical records, the Ming ambassador Chen Cheng reached Herat in 1414 by an overland route. Yet the Timurid empire did not reciprocate. Not long after Zheng He arrived, the Timurid empire sent a trade delegation comprising several hundred people to China in 1420, made up of mercantile groups from Herat, Bukhara and Samarkand. The delegation reached Nanjing. Whether or not the formation of this group has any relation to the arrival of Zheng He is a question that deserves further investigation by historians.

Among the materials now extant that can provide at least some evidence is a poem written by Zheng He's assistant, Ma Huan (a Hui Muslim from Ningbo, who understood Arabic) on his return to China. The poem reads:

> *Hormuz adjoins the ocean's edge*
> *From Ferghana to Misr ride traveling merchants*
> *I have heard of how Wide-Gaze went as ambassador to*
> *remote lands*
> *How can that compare to the glory of our present*
> *Broad Benevolence?*

Ferghana was the name given in the Han dynasty to the Eastern part of present-day Uzbekhistan, i.e., the area around Samarkhand. Misr refers to present-day Egypt. "The Marquis of the Wide Gaze" was the honorary title of Zhang Qian, the Han dynasty ambassador to the Western Regions. Ma Huan compares his and Zheng He's mission to that of Zhang Qian, and professes that the title of "Marquis of the Broad Gaze" cannot be more splendid than his present-day "Broad Benevolence."

In Zheng He's fleet, a number of seamen and translators were Muslims. In this regard, Zheng He's own family deserves mention.

Genghis Khan's first western conquest was undertaken to launch a punitive attack on Khwarezm, which was at that time holding sway in Central Asia. When his armies attacked Bukhara, the local chieftain Sayyid Ajjal Shams al-Din (Sayyid Ajjal meaning "sacred descendant") was only ten years old, and under the direction of his mother, led his hordes and opened up the city, offering his services to the Mongol army. Later, Sayyid Ajjal Shams al-Din followed Genghis Khan on conquests of different places. Subsequently, he further served Monge, a grandson of Genghis Khan and later Emperor Xianzong of Yuan Dynasty. When Monge's brother Kublai became emperor, Sayyid Ajjal Shams al-Din served as an official in various parts of China, finally in his old age being dispatched to Yunnan to serve as a "Manager of Government Affairs" (i.e. provincial head), where he dealt with the complex ethnic problems of the region. After his death, Sayyid Ajjal Shams al-Din was given the posthumous title of Prince of Xianyang, and his descendants for the most part remained in Yunnan, becoming a part of the Hui minority there.

Zheng He was originally surnamed Ma, and both his grandfather and father had visited Mecca. At age eleven, he was taken prisoner by the Ming army, and subsequently followed the Zhu Di, Prince of Yan, and later Emperor Yongle, on his conquests, gaining his profound trust. After

Zhu Di had acceded to the throne, he gave Zheng He the surname "Zheng." Zheng's genealogy explains that he was a sixth-generation descendant of Sayyid Ajjal Shams al-Din. His ancestor six generations before had come from Central Asia to make his home in China, and he had also gone to Central Asia to carry out diplomacy in the capacity of a Chinese official at the Emperor's behest, a story which was to become famous in Chinese history.

Anatolia, the Western Arabian Peninsula, and Egypt

After the Mongols' western conquest, the Ottoman Turks rose in Anatolia and gradually replaced the long-since Persianized Seljuk Turks. By the start of the fifteenth century, the Ottomans had already entered the Balkan peninsula, and had conquered Serbia, effectively encircling the Byzantine empire. Yet their heyday was still to come.

At that time, the Ottoman Sultanate was bordered in the South by the Mamluk kingdom (1250–1517), which was made up of slave-troops. In 1260, the Mamluk army first successfully attacked the Mongols, then defeated the Crusaders and annihilated the remnants Crusader forces, saving Islamic culture. They rebuilt the Egyptian empire, at the same time controlling the trade routes to Syria and the pilgrimage routes to Mecca and Medina, gaining abundant revenues.

Cairo was the capital of the Mamluk kingdom, and had always played an important role as a centre of Islamic culture. The famous historian Ibn Khaldun (1332–1406), at the time of Zheng He's voyages, was lecturing and writing at the Al-Azhar Mosque in Cairo. He studied the history of various different human civilizations and kingdoms, proposing a law which stated that all civilizations and dynasties had a cyclical rise and fall. This cyclical view of history was to influence the twentieth century historian Toynbee, leading the latter to conclude that the twenty-first century should belong to the Asian peoples of the Asia-Pacific region.

The Byzantine Empire and Russia

The rulers of the Byzantine Empire, which used Greek and was situated in Constantinople, still referred to themselves as Romans, even after the split with the Roman Latin Church in the mid-eleventh century. When Zheng He went on his voyages West, the Byzantine Empire was on the brink of collapse, and all that remained was the area around Constantinople. Even the city of Constantinople itself had sections under the control of Venice and Genoa.

Not long before this, the Byzantine Emperor had been prepared to unite with the Latin Church, but the church patriarchs and clergymen of Constantinople were unwilling to give up their own orthodox religious teaching and authority and be on an equal footing with the clergymen of the Latin Church. They would rather be ruled by the Ottoman and Islamic traditions, because in that way it would by contrast guarantee that the Greek Orthodox church patriarchs and clergymen could manage the internal affairs of the Eastern Orthodox communities.

Precisely because Constantinople was in a perilous position, the Russian Orthodox Church in the North subsequently broke away from the church patriarchs of Constantinople and became an independent Russian

Orthodox Church. Politically, the Grand Principality of Moscow was gradually ascendant and during the time of Grand Prince Vassily II (1425–1462) suppressed the power of the Mongol Kipchak Khanate and the control of the Tartars (the appellation used by the Russians to refer to Islamicized Mongols; the term here has a different meaning than when used to refer to the Tartars in the North of China). They also annexed several neighboring principalities, paving the way for Ivan the Great (1440–1505), who was soon to enter the stage of history. Russia replaced Byzantium as a political and cultural entity, and began to position itself as the successor to the Roman Empire, styling itself the "Third Rome". The Russian word "Czar", for instance, is derived from the Roman title "Caesar". Clearly, the Russian political system and the arts were all influenced by Byzantium.

Italy

In the fifth century, when northern barbarians invaded Rome, Europe entered the Dark Ages. Nine-hundred years later, European civilization was to shine again. And it could not be more appropriate that this period of "rebirth," the Renaissance, began in Italy. At the end of the fourteenth century, the Catholic Church split, with a pope in both Italy and France, and thus the power of the Vatican in Italy was weakened and the Holy Roman Empire made up mainly by the Germanic peoples and given its title by the Vatican, remained mostly in name. People began to have new room for activity. In Florence, the weaving industry developed, and an early form of capitalist production relations emerged, banking and financial services began, and rich merchants controlled city-states. In Venice, maritime trade brought it huge riches, and it also controlled some colonies in the East of the Mediterranean. The city states that broke free of aristocratic feudal rule formed certain political alliances.

In such an era of increasing wealth and gradual dissemination of knowledge, humanism began to sprout in Italy. Certain thinkers and writers in the fourteenth century revealed a way of thinking that placed man at the center, opposing the human against the divine, and challenging divine authority. Most importantly, this enlightenment thinking movement opposed obscurantism and mysticism, thus triggering an extraordinary advance in human understanding of the natural world.

In this way, then, fifteenth-century Italian crafts, carving, painting and architecture reached an unprecedented new height, whilst literary and philosophical innovation became the underlying basis of Renaissance thought, leaving a crucial legacy for ages to come.

At this time, Zheng He may or may not have known that the Europeans, especially the Italians, were building up their strength and preparing to soar forth to new heights. And the Europeans, including the widely knowledgeable Venetians, certainly would not have known that Zheng He in far-off China was engaging in the largest-scale voyage ever seen in human history

The German Region

The German region, which had always been relatively backward, was in name led by the Holy Roman Emperor, but a political system possessing actual administrative power had yet to be established. The politics of the German region consisted largely in the struggle between burghers and feudal lords. The Hanseatic League, formed of guild-like organizations known as Hanse, became a trans-regional binding force. However, by the fifteenth century, the League's gatherings had gradually decreased, and it eventually declined and was abandoned.

From the above brief description, it is perhaps difficult to discern that the innovation with the biggest motivating power for the European Enlightenment and the religious revolution in the sixteenth century that followed, actually came from the German area of the Rhineland. This was Gutenburg's movable type method (first used around 1450, it came some four hundred years after Bi Sheng's invention of movable type in the Song dynasty). The improvement of this printing technology allowed users of alphabetic scripts to set and print large amounts of books and materials quickly, thus hugely accelerating the dissemination of knowledge.

Northern and Western Europe

At the turn of the fourteenth and fifteenth centuries, Western and Northern Europe had entered the final period of feudal society. Agricultural production saw a relatively quick development, whilst industry (especially wool spinning), commerce, trade and the banking industry had all entered a new era. The merchants, with their newly accumulated wealth, had conflicts both with the handcraft industries and with the feudal aristocrats. The mechanisms by which order was maintained in the feudal system – self-sufficient production for the manor and the vassal relations between large and small feudal lords – were dealt a serious blow.

In Northern Europe, Denmark, Sweden and Norway

shared close relations both ethnically and linguistically, and in their customs. At the end of the fourteenth century, by way of several royal marriage alliances and diplomatic appointments, the Swedish Erik of Pomerania was jointly crowned king of Denmark, Sweden and Norway at Kalmar in 1397. These three countries thus united to form a Scandinavian alliance. Because of their geographical position, the three countries had all along received little control from the Vatican, and thus in their ideology had been subject to relatively few of the restraints of dogmatism of the middle ages. However, because of the relatively slow development of northern Europe, scientific innovation and humanist philosophy were not as advanced as Italy, England and France. The warm breeze of the Renaissance would take another hundred years before properly reaching cold Northern Europe.

The situation in Flanders (present-day Holland and Belgium), however, was not the same. Flanders was rich. It imported wool from England, and then exported it back to England in the form of woolen textiles, yielding considerable profits. The textiles industry together with the original agricultural industry drove the economic development of Flanders: its different handcraft industries, commerce and the banking industry all began to flourish, and the social power of the cities' inhabitants grew ever greater. Precisely because of such riches and social changes, many wealthy people would commission artists to paint pictures for them, and thus fifteenth-century Flanders saw the emergence of a large batch of outstanding painters. Many of their paintings took as their theme people's daily life, and were no longer limited to religious subjects.

However, in respect of politics, Flanders was a vassal state of France and its feudal lords pledged their loyalty to the French king. During the Anglo-French wars of the fourteenth century, English inhabitants in the country were actually arrested.

When Zheng He was making his voyages, England and France were engaged in the Hundred Years War. The war had begun as a result of a dispute between English and French leading aristocrats as to the right of succession of the French king. A further factor was a dispute over a large area of rich territory in the South-West of France.

Inspired by the martyrdom (1431) of Joan of Arc, the French army finally retook the land lost, defeating the invading English army. In fact, whoever won or lost the Hundred Years War, it was the death-knell for the old feudal system in Europe. In the space of a hundred years, both England and France, through changes within society and the needs of a foreign war, had crossed from a locally divided aristocratic feudal system to a nation-state in which the monarch possessed actual power. From this time on, the English court no longer used French. Both countries' kings established cabinets and state governmental mechanisms, thus initiating a situation which involved not simply possession in name but actual control of the state. In another respect, in complete contrast to the eleventh and twelfth centuries, under the national monarchy, the power of the ruler increased whilst that of God decreased.

The formation of nation-states governed by monarchs brought about advancements in the various European countries in regard to concentration of resources, integration of social forces and unification of popular mentalities. Subsequent European colonial expansion was likewise chiefly realized under the system of national monarchy.

When one considers the development of humanity as whole, China, situated at the Eastern extreme of the Eurasian land-mass, is an example of an early-maturing model. Already in the second century B.C.E., it had begun an unbroken system of absolute monarchy that by Zheng He's day had become an unalterable truth for any Chinese. And yet they could not know that at that time, in far-off countries such as England, France, Spain and Portugal, an absolute monarchy had only just emerged.

Spain and Portugal

In 711, the North African Muslims crossed the Straits of Gibraltar and entered Europe, sweeping across the entire Iberian peninsula. In 732, at Poitier in the mid-west of France, they were defeated by the grandfather of Charlemagne, and subsequently retreated south of the Pyrenees. For three hundred years after this, the whole Iberian peninsula, with the exception of a small corner in the North-West, fell within the scope of Islamic power. However, from the eleventh century, Christians separated into a few small kingdoms and over the course of five hundred years, slowly "reconquered" the Iberian peninsular from North to South.

At the time of Zheng He's voyages, Iberia was essentially divided into four parts. In the South-West was Portugal, controlled by the court of Aviz. The largest and strongest part of the peninsula was the kingdom of Castille. To the East was the kingdom of Aragon, which also controlled Sicily.

At the South of the peninsula was the Nasrid dynasty, established by the Muslims, who had been forced to retreat that far; this was also the last Muslim regime seen on the Iberian peninsula. Although its territory and power had become far less than the former Muslim regimes, it was still able to maintain its economic prosperity and religious tolerance. At the start of the fifteenth century, the Nasrid dynasty built a palace that has become particularly precious in the history of Islamic art – the Alhambra.

The Castillian queen Isabella married the Aragon king Ferdinand, uniting the two countries, an event which happened half a century after the death of Zheng He. In 1492, unified Spain destroyed the Nasrid court, banishing Islamic power for ever from the Iberian peninsula. In the same year, Columbus, the Italian whom they sponsored, led three sailing ships to the other side of the Atlantic, to the shores of Central America.

While Spain certainly occupied a number of places in the Americas first, owing to its sponsorship of Columbus, the state that really put its efforts into developing maritime activity, and which did so in a sustained fashion for one hundred years, was Portugal, situated on the edge of the Atlantic. Due to their isolated position on the West of the Iberian peninsula, the Portuguese had no way to participate directly in the lucrative Mediterranean trade, and so having expelled the Muslims, they made use of Muslim seafarers' navigational technology and geographical knowledge, and gradually pushed south along the West African coast, with the aim of finding another trade route East. In round 1430 – the same time that Zheng He made his last voyage, to Kenya on the East African coast – the Portuguese occupied the Azores, before arriving at Cape Bojador on the West African coast.

The arrival of the Chinese fleet on the East African coast at the same time as the European Portuguese arrived on the West African coast could be said to be history's first accidental, yet deeply significant instance of "East and West reflecting each other."

However, after Zheng He's seventh voyage, i.e. after having visited Calicut on the South-West edge of India for the seventh time, Chinese ships did not appear in the Indian

Ocean again. The Portuguese continued to grope their way forward, and in 1460 reached Sierra Leone. In 1498, da Gama rounded the Cape of Good Hope, the southernmost tip of Africa, and forged into the Indian Ocean, reaching Zheng He's "second homeland," Calicut.

From this time on, the Indian Ocean's waves were to become a place where European fleets would hold sway. The Portuguese fleet in 1511 occupied Malacca, and in 1517 reached the mouth of the Pearl River; in 1543, they reached Japan. The curtain had been raised on the era when Europe was to rule the world.

(May 2005)

9

From Movable Type Printing to the World Wide Web

The Northern Song Dynasty: A Period of Flourishing Culture

The Northern Song Dynasty (960–1127) is a period in which Chinese culture was approaching maturity and exhibiting rich creativity. The studies of the Confucian classics, history and philosophy were prospering, and the arts of poetry, prose writings, calligraphy and painting were marching forward in big steps. The civil service examination system was near perfection, and public and private academies were showing signs of great vitality.

The many scholars from the mid-eleventh century of the Northern Song period have left behind a precious spiritual treasure, including these lines: "To worry long before others get worried, to rejoice only after others have rejoiced " by Fan Zhongyan, "The blossoms fell, not heeding our human wish" by Yan Shu, "The drunkard's mind is not on the wine" by Ouyang Xiu, "Parting with a loved one has ever been painful since days of old" by Liu Yong, "The call of spring distracts me, robbing me of my sleep" by Wang Anshi, and these gems

Based on a lecture delivered at the inauguration of the Cultural Lecture Series organized by the Chinese Civilization Center at City University of Hong Kong. First published in *Hong Kong Economic Journal Monthly*, December 1988; reprinted in *Essays on the Humanities from Universities in China*, Vol. 3, June 1999.

by Su Shi, well-known for his calligraphy, painting, poetry and essays: "I only hope we may have long lives and share the moon's beauty, though a thousand miles apart," "Ten years – dead and living draw apart," "Where in the world cannot one find fragrant grass?" and "The true face of Mount Lu is hidden from us." All these have become part of our day-to-day vocabulary.

Covering a period of more than one thousand years, the several hundred *tomes* of *The Comprehensive Mirror to Aid in Government* compiled by Sima Guang are the most important historical treatise written in China in the last one thousand years. The brothers Cheng Hao and Cheng Yi, for their part, laid the foundation of the Confucian thoughts for the millennium to come with their moral philosophy. Yet, I am afraid that the sum total of the achievements of all these figures cannot measure up to the enormous and far-reaching influence that one obscure commoner produced on the human race. The person in question is Bi Sheng, the inventor of the movable type printing.

The Technology of Printing Spurs the Development of Culture and Education

According to *Dream Pool Jottings* by Shen Kuo (1031–1095) of the Northern Song, Bi Sheng invented the movable type printing at around 1045 C.E. He carved individual Chinese characters on types made of clay, and arranged them by rhyme-groups for easy retrieval. To prepare for printing, one has to first cover a metal plate with wax, put a rectangular metal frame on it, and arrange the types inside. When the types are set, heat the plate over fire. As soon as the wax

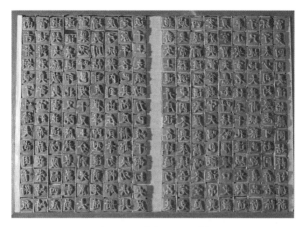

Reproduction of clay movable type

begins to melt, press down on the types with a board. The movable types are now affixed to form a page of text. Ink is then applied for printing. When the printing is done, the board is again heated to melt the wax away, and the types can be removed and stored for later use. Shen Kuo had this to say about movable type printing, "It is by no means easy to print two or three copies with this method, but if one is printing several hundred or thousand copies, the movable type printing is extremely fast."

Woodcut printing was widely used before the invention of movable type printing, and continued to enjoy extensive application even afterwards. This is done by carving text and pictures on a piece of smooth hard wood of dense grain. One block of wood is needed for each page. Ink is applied to the carved wood, which is then pressed on a piece of paper in the way one would with a seal. Woodblock printing began during the transition between Sui and Tang (circa 600 C.E.) *An Illustrated History of Printing in Ancient China*, recently published by the City University of Hong Kong Press and the

Cultural Relics Publishing House in Beijing, has a detailed introduction to the invention and development of printing technology, complete with text and illustrations. The prerequisites of printing are, of course, language, paper, ink and wood carving technology. Chinese language has an early beginning, and was standardized by the Qin dynasty (221–206 B.C.E.). Paper-making technology began in Western Han, but paper of high quality had to wait until Cai Lun of Eastern Han, who used fiber from plants to make paper in 105 C.E. By the Jin dynasty (circa 300 C.E.), paper had replaced bamboo slips and cloth as materials for making books. With its much lower cost than that of bamboo slips and cloth, paper facilitated the dissemination of knowledge. The invention of printing made it possible for knowledge to spread even farther, and the development of culture to speed up even more.

Printing technology was initially used for the printing of Buddhist scriptures and portraits of Buddha. Only later was it extended to the reproduction of Confucian classics, history texts, essays, *shi* and *ci* poetry and other kinds of writing. By the late Tang dynasty, bookstores began to appear in cities, and culture was no longer the monopoly of the nobility. The Song government established the imperial academy (similar to a public university nowadays), which was also in charge of book publications. In the seventeen years from 988 C.E. to 1005 C.E., the printing woodblocks that were stored in the imperial academy increased from 4,000 to 100,000. That the Song dynasty saw an abundance of talented individuals and enjoyed a flourishing culture cannot be separated from the rapid development of printing technology.

In addition to government publications (that is, those of the imperial academy), books were also printed commercially by bookstores and privately by individuals who financed the publication. Research shows that Jiangsu, Zhejiang, Fujian,

Jianxi and Sichuan were the places where the printing industry was most developed during the Song dynasty. Eighty percent of the books published at that time came from these five regions. Twenty-four thousand of close to the thirty thousand identifiable metropolitan graduates (*jinshi*) (that is, 80 percent) also came from these five regions. Measured by time, the number of metropolitan graduates at the end of the Song dynasty was four times that of the beginning, clearly showing the advanced development of economics and culture and the spread of education during the Song dynasty.

The Developed Industry and Commerce and the Advanced Technology of Song

The three hundred years of the Northern and Southern Song period saw the rapid development of the Chinese society. The population of Northern Song was about 80,000,000. Industry and commerce gradually developed, while the transportation system ran smoothly, and, thanks to printing, paper currency was becoming popular. Kaifeng, the capital, had an area three times that of Rome, and a population of well over a million. (The biggest city in Europe at the time was Constantinople, which had some 100,000 people.) By the eleventh century, China had begun to create high-quality iron with coal. The overall output of cast iron reached 114,000 tons (The British iron output in the eighteenth century was only half of this amount.) With the abundance of iron, agricultural implements underwent major improvement, and weaponry and armor were getting better by the day. Ten years after the unification under the Song dynasty, the whole country functioned as a unified internal market, and the commodity economy began to prosper. At the same time, there was a notable increase in foreign trade. Boats from the Persian and the Arab world arrived continuously at Yangzhou, Hangzhou, Quanzhou, Guangzhou and other places, carving out a Silk Route by water. Some boats built in the Song dynasty had as many as five or six decks, and over ten sails. They made use of the stern post rudder, navigation maps and the compass in sailing, and could carry five hundred people. (The big wooden crafts in the Mediterranean at the time were still relying on human labor and sculls.) With its cultural and economic power as well as its advanced navigation technology and weaponry, it is conceivable that China could have entered the age of discovery before Europe and set up colonies over the world. However, history did not develop along this trajectory. John K. Fairbank, an expert on Chinese history in the U.S., believes that China did not take part in any colonial enterprise not because it lacked the ability to do so, but because there was no motivation for it. This point was amply illustrated when a ban on ocean expeditions was declared after Zheng He's seven missions to the Western Ocean.

The Rise of Europe

At the time when Cai Lun was experimenting with paper-making technology, the Roman Empire was at height of its power around the Mediterranean. In the fifth century C.E., the Roman Empire came to an end in the hands of the barbarians from the north. Europe entered a stage of darkness. In 900 C.E., paper-making technology reached Cairo in Egypt by way of Baghdad, and made its way to Morocco in 1100 C.E. Spain, the major part of which had been occupied by Muslims from northern African since the beginning of the eighteenth century, got the printing technology from Morocco in 1150 C.E., and built the first paper mill in Europe, more than one thousand years after Cai Lun.

Woodblock printing technology was imported to Italy from China in the fourteenth century, probably by way of Persia and Turkey. The earliest printing product in Italy that we now know of was dated seven hundred years later than that of China. As for the movable type printing invented by Bi Sheng, similar technology emerged in Europe four hundred years later, which was used to print the *Gutenberg Bible* in Germany in 1455.

The arrival in Europe of paper-making and printing technology to a large measure stimulated and propelled the Enlightenment and the Reformation. Martin Luther described printing as the supreme gift from God, which enabled the gospel to spread far and wide. With the help of publications several dozen times cheaper than hand-copied manuscripts, European civilization took off and thrived. That the languages in Europe are alphabetical makes the application of movable type printing even easier, and the impact and influences are thus all the more obvious. Books on religion, law and technology proliferated, in turn causing the rapid rise in number of universities in Europe. At the same time, the movable type printing also encouraged the growth of popular literature, which not only promoted the use of the vernacular, but also indirectly advanced the cause of nationalism and prompted the birth of modern nation-states.

Western Europe entered the colonial period after the sixteenth century, giving an extra boost to its internal economic prosperity. Accompanying such economic rise is the emergence of modern science, rationalism and humanist thoughts. Isaac Newton of the seventeenth century is a major representative figure of modern European civilization. In mid-eighteenth century, James Watt invented the steam engine, sparking off the Industrial Revolution which would significantly increase human productivity. After that, Europe dominated the world. There is not a place in the world nowadays that has not come under the important influence of European civilization in materialistic culture, social systems and humanistic thoughts.

The twentieth century was incontrovertibly the American century. In addition to the airplane, the television and the atomic bomb, all of which are American inventions, the U.S. also produced the first computer in 1946, bringing the whole human race to the digital era.

The Digital Revolution
and the World Wide Web

The progress in quantum physics brought to the world the semiconductor transistor in 1956. By the beginning of the 1960s, transistors had gradually taken the place of vacuum tubes, which had been in use for the major part of the century. When I was a graduate student at Stanford in 1963, a fellow-doctorate candidate in electric engineering told me that they were researching a new technology which would produce by means of chemical etching an integrated circuit of more than twenty transistors on a silicon chip. At the time, I was not able to appreciate the significance of the technology, let alone estimate its impact.

In 1971, Intel produced the 4-bit 4004 processor, with approximately 2,000 transistors. Texas Instrument also came up with a similar product. I remember that I spent US$398, about one-third of my pre-tax monthly salary, on such an electronic calculator that could take the place of the slide-rule. Subsequently, Intel put in the market the 8-bit 8008 micro-processor in 1974, lending force to the growth of the micro-computer that was beginning at the time. The 286, 386, 386, and 586 (i.e., Pentium) processors came to be produced one by one, to the point that they have now become common knowledge. The latest Pentium II micro-processor indeed deserves the name of "micro": on a silicon chip of less than 0.5 square millimeter are engraved an extremely complicated circuit made up of eight million transistors. The distance between two electric currents is only 0.2 microns, or 0.0000002 meter.

From the end of the 1960s to now, the computational or memory capacity of the silicon chip increase two-fold every eighteen to twenty-four months, and the price of the chip keeps falling rather than rising.

The digital revolution was of course not limited to the improvement on the micro-processor or the progress in computer hardware alone. In 1971, some people in the U.S. connected twenty-three computers in accordance with the protocol known as TCP/IP to form a network and called it the Internet. After more than twenty years of development, especially after the rise of micro-computer, the Internet has become the most comprehensive, exhaustive and far-reaching network of information exchange in the world. By means of fiber-optic cables, users of the internet can easily utilize the functions it provides such as email, FTP (File Transfer Protocol), electronic commerce, and video-conferencing. There are approximately twenty million computers connected to the Internet in the world today, and more than one billion people using it. The amount of information traffic on the Internet doubles approximately every one hundred days.

In 1994, World Wide Web, made possible by the

hypertext transfer protocol (http), came to the Internet. After four years of speedy development, there are now approximately five million websites, all of which are easily accessible through search engines. A website can accommodate text, graphics and sound. The latest micro-processor also makes the use of three-dimensional multimedia products even easier.

Managing Chinese Electronic Information

Chinese is made up of mono-syllabic characters, which combine to form words. There is a high degree of flexibility in the composition of Chinese language. For reasons of rhetoric and inflection, Chinese language offers great freedom in the choice of grammar and lexicon features. All these give Chinese literary works, especially in the poetic genres, a unique kind of flavor and cadence. However, as it is made up of individual characters that evolved from ideographs, Chinese language poses immense difficulties in writing. In modern times, it presents significantly more problems than phonetic languages in typing and in printing. Chinese culture and education have obviously fallen behind Europe and North American as a result, and the task of bringing them to a wider public is rather difficult. For that reason, ever since the May Fourth Movement, many Chinese intellectuals have found their emotions and rationality at odds with each other. On the one hand, they believe that like most people in the world, they have to adopt a phonetic language in order to make education and culture accessible to the public so as to modernize China. On the other hand, they have such deep feelings towards the highly artistic written language left behind by their ancestors that they are unwilling to give it up.

Yet, precisely because Chinese written language is made up of individual characters and not spelled out phonetically, the invention of printing and the increased circulation of books have afforded our highly populous and linguistically diverse country with a huge body of common texts. These texts constitute an invisible but powerful congealing force to our culture not found in any other group in the world. For this reason, various governments since the May Fourth period have either tried to supplement the traditional written language with a set of phonetic symbols or simplify the written language. None, however, has shown the resolve like Korea or Vietnam to make Chinese writing completely phonetic.

Another line of Song poetry comes to mind, "Just when we seem to have come to a dead end with layers of hills and crisscrossing waters ahead of us, the horizon opens up, beyond the flowers and the willows, and we are faced with a new world." At the end of the 1970s, the Chinese government unveiled the second round of simplified Chinese characters. As they were simplified to such a degree that they did not resemble Chinese any more, they were rejected by the majority of the population and were given up at the end. Precisely at this time, micro-computer came to the scene. The cost of electronic storage came down drastically, and the talented among the Chinese living in the Mainland, Taiwan, Hong Kong and North America designed a number of methods of inputting Chinese, which made the typing of Chinese almost as fast as any phonetic language. Later, researchers at Peking University invented the method of laser typesetting, greatly increasing the efficiency of printing Chinese. The difficult problem that had plagued many Chinese intellectuals for almost one hundred years was thus solved. With the development of micro-electronic and computer technology, the management of electronic information in Chinese is no longer an obstacle to the development of Chinese culture. The movable type printing invented by our ancestors provided Europeans with their phonetic languages a perfect chance to popularize their culture; and now the computer and laser technology from the Europeans and Americans provides us with our ideographic language a new impetus to popularize ours. The world is fair, after all. We with our ancient history and distinctive cultural tradition are now facing a new prospect.

Chinese Culture in the Age of the World Wide Web

Computer, the Internet, email, and especially the popularization of the World Wide Web allow the dissemination and exchange of information in Chinese to be done with the same ease and speed as any phonetic language. The problem that we are facing now is how we can make use of the new technology to enrich the content of Chinese electronic information.

Chinese of the twenty-first century should learn how to race comfortably down the information highway that is getting wider by the day. We should not be contented with using only the pre-existent highway either, but build more and better highways of our own. We should also build a large number of "cars" with engines made of the Chinese language. Recently, the City University of Hong Kong established the School of Creative Media, with the hope of cultivating talents in this area. We should aim at creating in the post-digital revolution age of the World Wide Web a new civilization that is more glorious than that of the Song dynasty which had brought us the movable type printing. I call this enormous project the "C++ project." "C" refers originally to the computer language that became popular about ten years ago. After due improvement, it became the "C++" language that has come to wide use. I am borrowing this term of information technology here to lend to the project the appropriate aura of our age, and also to underscore the enormity and complexity of the task of creating the new culture of the age of the World Wide Web.

In the way we use it, the "C" stands for Chinese culture. It goes without saying that the new culture that we want to create for the age of the World Wide Web is based upon our traditional culture. The first "+" sign indicates that we will not accept the cultural heritage left to us by our ancestors without due selection. Rather, we have to engage in deep and critical reflection on Chinese history and culture. For example, even though the Song dynasty can be shown to have reached a certain degree of maturity in the areas of politics, economics and culture, why is it that our society as a whole did not take off in the way that Europe did, which ultimately led them to the Enlightenment? Or, for all the advancement and creativity that Chinese craft and technology had demonstrated since the Song dynasty, why did we never develop modern science based on mathematical deduction and repeated experiments? Furthermore, even though humanistic thoughts and the social ideal that "people are most important and ruler least" was articulated very early in China, why is it that the kind of European humanistic and democratic ideas did not appear in modern China?

The second "+" sign points out the need for us to be open-minded and be equipped with an international vision. We should learn from our ancients of the Wei, Jin, Tang and

Song dynasties – confident and open, and be ready to absorb the beneficial elements from foreign culture. Buddhism, which we have by now sufficiently digested, came from India. Chinese painting, music, dance, sculpture, architecture and even food and clothing have all come under the influence of India and Central Asia. We might have already forgotten that trousers, the chair, the lute, and even the flat bread that we eat have all originated from the West.

We now find ourselves in a highly developed informational world, where economics and culture are gradually getting global in scope. Living in Hong Kong, the most modern, the most experienced in international exchange, the freest and the most open place in China, we are in the position to give a bit more to the "C++ project." Only by revitalizing the entire Chinese people and contributing once again to human civilization in the next century can we face our ancestors and our descendants with pride.

10

From Geometry to Tap Dance

When I first came to work in Hong Kong in 1990, after having lived in Taiwan and North America, I immediately became comfortable with living on Chinese soil with a population that is 97 percent Chinese, but working in two cultures.

When I came to work for City University of Hong Kong six years later, I had already perceived the rich opportunities Hong Kong offered to scholars interested in cross-cultural studies. I had also seen the potential for Hong Kong to become a center for cultural exchange in the world. However, it was not until Professor Zhang Longxi joined City University in 1998 that I saw the feasibility of establishing a Center for Cross-Cultural Studies.

Now I am very pleased to be able to write down a few thoughts for the inaugural issue of *Ex/Change*, the newsletter of the Center for Cross-Cultural Studies.

Perhaps a good place to start is the introduction of Euclidean geometry to China by Matteo Ricci (1552–1610) through his friend and an early convert to Catholicism, Xu Guangqi (1562–1633). It was through this introduction that Chinese scholars gradually began to understand European philosophy and science based on the time-space framework and the logic derived from such a framework. Before that time,

Published in *Ex/Change*, No. 1 (June 2001), Center for Cross-Cultural Studies, City University of Hong Kong.

Chinese philosophers had focused on a time-state framework (e.g., birth, death, harmony, disorder), paying their attention mostly to systems rather than discrete components as was the case in European philosophy and science. It is now clear that the time-state framework is not so conducive to the development of modern science, the key to which is precisely the understanding of the temporal-spatial relationship and its logical deductions. However, modern science (medicine, space technology, etc.) has developed to such an advanced stage that the analytical skills applied to small components must be complemented by the knowledge of systems through synthesis and the holistic approach adopted by earlier Chinese philosophers.

I have devoted quite a few years to the study of fluid dynamics, including the physical phenomenon of diffusion, a process of exchange of particles (e.g., molecules) in a well-defined time-space framework. The study of this process cannot be done without a good knowledge of geometry and other mathematical tools. Thus, in a strict sense, the acquisition of modern science by the Chinese can be traced to the personal contact and exchange between Matteo Ricci and Xu Guangqi.

In an extended and perhaps metaphoric sense, the movement of people in time and space forms the basis of diffusion of ideas and technologies. Even though collisions occur at both the microscopic "molecular" level as in the case of diffusion as well as the more general or macroscopic level as in the case of cultures, the result of these collisions is always exchange and/or fusion.

The establishment of the Center for Cross-Cultural Studies is dedicated to the notion of cultural exchange and diffusion, and perhaps eventual fusion between cultures of different origins. The Center is a scholarly unit for the advancement of learning, but its work need not be beyond the understanding of ordinary people.

Although I no longer work on fluid dynamics, I have become more and more an enthusiast for cultural dynamics and cross-cultural exchange. Again in a metaphoric way, I can perhaps express my wish for the Center by describing three non-scholarly events in Hong Kong, namely, three concerts I attended recently.

The first was in early March. A new piece entitled *Guan Shan Yue* (The Moon Over the Pass and the Mountains), written by a contemporary Chinese composer Zhao Jiping, was performed with four instruments – *sheng*, *pipa*, cello and tabla. The performers called themselves, symbolically, the Silk Road Ensemble. Yo-Yo Ma was the cellist.

The second concert took place in mid-March. It was one of a series of *erhu* concerts organized by the Hong Kong Chinese Orchestra, one of the largest and best Chinese orchestras anywhere. The program included a solo performance by a well-known *erhu* musician of *Moto Perpetuo* by Niccolò Paganini.

The third concert was given by the Hong Kong Philharmonic Orchestra in April in a program called Dances Around the World. Under the baton of Dr. Samuel Wong, a Hong Kong native, this flagship orchestra of Hong Kong played a novel piece by an American composer called *Tap Dance Concerto*. The music was very rhythmic and somewhat jazzy but definitely classical in format. A Chinese tap dancer played the part of the soloist in the concerto, even giving an impressive cadenza in the second movement with his amazingly nimble feet.

The three concerts are typical of the arts and culture activities that most often combine the best of the Chinese and the Western in a single performance, creating an artistic harmony of diverse themes and talents from the East and the West. And that is Hong Kong. Could there be a better place for the establishment of the Center for Cross-Cultural Studies? I hope all those who are interested in the exchange of cultures and would like to see fusion of cultures taking place will join me in wishing the Center success and supporting its activities, which, as the title of its newsletter indicates, are all about cross-cultural *Ex/Change*.

11

A Dose of Idealism

Hong Kong, without doubt, is heading into challenging times. Yet we should not be overwhelmed by the dismal news from day to day, however depressing it may be. We have to remember this international city, which most of us call home, is unique in China and probably the world. Nowhere else can we find pragmatism and idealism meeting in such an extraordinary way, often with dramatic results, in all realms of life – political, economic, social and cultural. Let me explain.

No one doubts Hong Kong is a city in China, and, by all reckoning, its most cosmopolitan. The idealism of the one country principle is matched by the pragmatism of the two systems arrangement, under which China and Hong Kong have their own membership in the World Trade Organization and are counted as separate trade zones.

In sociological terms, pragmatism and idealism also flourish side by side in Hong Kong. Consider C. Y. Tung, the late shipping magnate, who is remembered mainly for his business acumen. Few know of his contributions to international understanding, such as the establishment of the Seawise Foundation to promote the concept of shipboard education. Even to this day, the University of Pittsburgh, where I worked before coming to Hong Kong, has a Seawise-affiliated program that takes six hundred students at a time

Published in *South China Morning Post*, March 3, 2003.

around the world on board ship to gain experience of foreign cultures.

The people of Hong Kong like money, but they mix a down-to-earth sensibility with healthy imaginations. Right now, with the economy on a downswing, music classes are thriving. Why? Because parents think it will help their children get into secondary schools or universities. If this is not a case of pragmatism and idealism meeting, what is? Culturally, the same quality of openness is pervasive. An important element of Chinese culture is its inclusiveness. While the majority of the population in Hong Kong is Chinese, the non-Chinese communities contribute to the city's pluralistic and international character. The beauty of Chinese culture is that we can embrace it along with an array of foreign cultures. We should encourage not only diversity but also the fusion of traditional and modern culture, and popular culture with refined culture. A Chinese saying sums this up perfectly: *The vastness of the ocean stems from the many rivers it embraces.*

Lastly, pragmatism and idealism meet in Hong Kong economically, I believe. We should take our inspiration from France, a country which honors its writers, artists, and philosophers but which has not forgotten that they can also bring income to the country. Hong Kong should promote cultural tourism to, among others, mainland tourists, who will come to see for themselves a Chinese city that has had a hundred and fifty years of close contact with Europe. At the Culture and Heritage Commission, senior business leaders bring business sense as well as their enthusiasm to projects that enhance the city's cultural life and also make good economic sense. In this regard, I am glad the Hong Kong government is also promoting the creative industries as a way to boost the economy.

Despite the city's economic woes, the government is still determined to open the West Kowloon reclamation area. The area will be designated an integrated cultural and entertainment district. It will attract tourists, elevate the quality of life for its citizens and, by promoting and selling culture, make a name for this vibrant, cosmopolitan city. What better example of the marriage of idealism and pragmatism that makes up Hong Kong's soul?

12
SAR's Version of Cultural Fusion

In a lush illustration of the way Eastern and Western cultures mix nowadays, we were treated on a recent Sunday to a front-page photograph of the "Three Tenors" – Jose Carreras, Placido Domingo and Luciano Pavarotti – taking a bow on a gigantic stage in front of the gate to Beijing's Forbidden City while surrounded by women dressed in richly colorful Peking-opera costumes.

The venue was Chinese and six hundred years old. The singing was European and part of a four-hundred-year-old tradition. It was truly an East-meets-West moment.

We in Hong Kong are experienced with such moments – after all, East meets West here every day. We take great pride in that fact. Even so, not everyone gives much thought to what East means or to what West means or even to what "meet" implies.

To most Hong Kong people, East presumably signifies people of Chinese ethnicity and customs from the Chinese tradition. Similarly, West probably means people of Western European descent (and, in our context, of primarily Anglo-Saxon stock) and the traditions they share. But when it comes to "meet," the meaning is much less certain.

We should, at a minimum, have respect for cultures other than our own. We should also engage in cultural exchange and endeavor to have cultural fusion. But many Hong Kong

Published in *South China Morning Post*, July 16, 2001.

residents – despite frequent contact with people of Indian, Nepalese or Pakistani background – have no knowledge of South Asian cultures. And some long-term, non-Chinese residents have not taken advantage of their presence in Hong Kong to learn about Chinese languages and culture.

We must do better than recite "East meets West" as a mantra.

The Culture and Heritage Commission, which I chair, is trying to do its part. In March, we published a consultation paper on cultural development. The consultation period expired at the end of last month. The document offers this vision: "It is our long-term goal to expand our global cultural vision on the foundation of Chinese culture, drawing on the essence of other cultures to develop Hong Kong into an international cultural metropolis known for its openness and pluralism."

Chinese culture as we know it today is the result of many centuries of evolution, during which Chinese traditions absorbed and integrated elements of foreign cultures.

During the Han dynasty (206 B.C.E. to 221 C.E.), the city of Loulan in today's Xinjiang province served as a key trading post along the Silk Road, a gateway to and from China and focal point for Central Asian, Chinese, Indian and Persian cultures. This role was supplanted and amplified from the fifth century onwards by another major station on the Silk Road: Dunhuang in today's Gansu province, home to famous grottos. Though located at China's geographic periphery during their heydays, both cities played important roles in the development of Chinese culture.

The best-known example of Chinese culture being influenced by foreign culture is the third-century introduction of Buddhism, via the Silk Road, and the religion's subsequent integration into Chinese society. There were also foreign influences in architecture, dance, music, painting and sculpture. A case in point is one of the most popular and essentially "Chinese" musical instruments, the *huqin* (or "barbarian fiddle"), a stringed instrument originating from Central Asia.

Several recent musical events in Hong Kong are worth noting. In March, American cellist Yo-Yo Ma and his Silk Road Ensemble performed using instruments from China, Europe and India. Also in March, the Hong Kong Chinese Orchestra gave a concert in which a *huqin* soloist played Moto Perpetuo ("Perpetual Motion"), written for the violin by Italian composer Niccolo Paganini. In April, the Hong Kong Philharmonic played Tap Dance Concerto, in which a local tap dancer played the solo using his nimble feet. Last month, the Hong Kong Sinfonietta performed a composition by Lo Wing-fai that similarly blended Chinese with Western.

Much like Loulan and Dunhuang in ancient times, Hong Kong is at the periphery of China, a gateway to and from the mainland, and an important international trade center. We are well positioned to serve as an international cultural center and to influence the cultural development of China as a whole.

If the recent Three Tenors concert symbolizes China's attempt to get acquainted with the best of European culture and to learn from it, the four concerts in Hong Kong suggest a process of fusion is taking place here.

Each of these five concerts testifies to China's quest for nutrients for its ancient culture. They are harbingers of the kind of Chinese culture – of which Hong Kong's is a special component – that might emerge in the new century.

13

On Models of Cultivating Creative Talent

"A closely tended flower will not bloom,
but a carelessly planted willow will grow into shade."
有心栽花花不发 无心插柳柳成荫

I will build this up piece by piece, like woodblocks. I will first talk about innovations, then innovative talent, then the cultivation of such talent, and finally models for the cultivation of such talent.

Thinking Outside the Box:
the Basic Tenets of Innovation

First of all, let us take a look at this painting from the sixteenth century, the period of Timur's grandchildren. It is a Persian miniature painting, and an illustration in a book. The unusual thing about this painting is that the artist put the tree and rocks outside of the frame of the box, which is smaller than the size of the page. Why did I pick this picture as my first slide? Because I think that fundamentally, innovation involves thinking outside the box. This artist has chosen to express his creativity by breaking the confines of the box.

Let me now introduce to you what to me are instances of innovation: the articulation of the relationship between the

---

Based on a speech the writer delivered at the International Forum of University Presidents organized by the Ministry of Education of China, held in Shanghai in July 2006.

axis and the wheel, or the making of paper and the wooden-recurved bow. The latter was invented by the people of the steppes in the fifth and sixth centuries A.D. Its strength was comparable to the considerably heavier and more cumbersome long bow of the twelfth and thirteenth centuries C.E. in England. The wooden recurved bow was an important innovation in the history of human civilization and warfare.

And then there are the stirrups. Also invented by the people of the steppes, they made the job of controlling horses much easier. Before then, with only reins, which made it necessary for the rider to use his hands to steer the horse, riding was far from simple. Stirrups, however, allow a good rider to steer the horse with only his legs, and for that reason, they have exerted a huge influence on transportation and warfare. Other examples of innovation are the printing technology of China, the rudder, gunpowder and compass, all of which are extremely important innovations in the development of human civilization.

Coming closer in time, examples of innovation include the invention of the steam engine, followed by that of the telegraph and the telephone. Besides these technological innovations, there have also been important breakthroughs in the natural sciences and mathematics.

An extremely important innovative concept in the progress of human thought, logic and mathematics, but one that has no counterpart in real life, is "zero". The Indians first came up with the concept, which was later borrowed by the Arabs. It is now known as one of the Arabic letters, and is represented by a dot. What an important dot (or "0") it is in mathematics! The invention of "0" was thus also a significant conceptual and practical innovation.

A further example: the binary system proposed by the German scholar Leibniz is made up of 0 and 1. Thanks to him, the operation of computers nowadays is based on the binary system. Another contribution of his was the development of calculus.

Yet other examples are Newton, who discovered the three principles of motion as well as gravity; Mme. Currie, who was responsible for the basic theory and experimentation of radioactive elements; and, of course, Einstein, who came up with the theory of relativity. Closest to us in time is the important discovery of Watson and Crick in 1953, who sketched for us the structure of DNA in the shape of the double helix. All research on bio-technology and genetics today is built upon our understanding of DNA. This was a ground-breaking discovery and a remarkable innovation.

Of course, innovations are not limited to the field of science and technology but are found in views and thought in the humanities, politics and economics. For example, the civil service examination system came into being during the Sui dynasty in China, and we have a modern version of it in the college entrance examination today. Another innovation is the "flying money" (*feiqian*), where a person could specify his credit on a piece of paper, stating where money could be drawn. Upon its receipt, the financial institution concerned would then issue the money. This was invented in China, as was paper currency, which was widely used during the Song dynasty, and gained even wider circulation in the Yuan dynasty.

There are a few other innovations that I would like to discuss. In the past, accounting was done in the single entry format. In the fifteenth century, a Venetian came up with the method of dividing an accounting sheet into two columns, one showing assets and credit, and the other liabilities and debits. Today's accounting practice is based on this method. If the single-entry method were still in use today, it would create a lot of confusion in the finances of many commercial interests and governments. The double-entry accounting method was a major innovation in human economic and financial activities. The Dutch East India Company also comes to mind, which first came up with the concept of stocks and bonds in 1602. Then there is Wall Street, a small street in New York City, but its influence is felt far and wide. In 1792, twenty-four stock agents on this street got together and wondered why they could not collaborate to represent the interests of their clients. This was the beginning of the world's earliest stock market – an important innovation, to be sure. From that point on, the center of world economics was beginning to shift to the United States.

All the examples I have raised are important innovations. All of them have one thing in common, namely, that when they were first proposed, they produced an impact on or even caused destruction to the economic activities and technology of their time. They would not be known as innovations otherwise.

The Four Ways
in Which Innovation Emerges

I shall now put the two words "innovation" and "talent" together. What is innovative talent? What do talented innovators have in common? I feel that people who have an understanding of natural or social phenomena and recognize their importance are deserving of the name. They do not necessarily have to invent something and come up with a new product, nor is it necessary to start everything from scratch. People are innovators if they can reorganize existing theories, scientific knowledge or materials, and then propose new insights, or come up with new products or new applications. As long as a person or a group of persons can subject existing things to reorganization and on the basis of which produce new functions, they are innovators. In addition, some are capable of finding connections between things that are usually regarded as unrelated. What they do amounts to innovation, too.

For example, on a cloudy day in the nineteenth century, a researcher of photography did not want to waste his film and so he put it away in a drawer to wait for better weather. A few days later, he found that the film had been exposed to light. An average person might do no more than find new film for the camera, but our photographer went on to ponder why film left in a dark drawer would still be exposed. The photographer in question was Roentgen, who discovered X-rays. He was an example of those who can discern a phenomenon in and find an explanation for things that are seemingly unrelated. I regard that as an innovation, unintended as it may have been in the beginning.

Perhaps, for some, the process of innovation is a matter of luck. For others, the same process may take a long time, but in the end, they also attain their goals. Moreover, they also recognize the importance of what they find. I am reminded of a famous line of poetry by Xin Qiji. You may be surprised by my interpretation of his words. Let us say that you have been looking for something for the longest time but still fail to find it, when, all of the sudden, it comes into view in a flash. Xin Qiji puts it this way,

"I looked and looked in vain for her in the crowd. Without thinking, I looked over my shoulder, and there she was, standing where the lights were far and few."

Now, the person that you are looking for may not mean much

to you, and you would not call her an innovation anyway, but the process that you have just gone through is an innovative process, particularly common in instances of literary inspiration.

Let me offer some preliminary conclusions. I think that there are four kinds of situations in which innovations are likely to succeed, be they scientific, material and systemic inventions.

The first is luck. A company is known to have tried to develop a kind of powerful glue. It conducted a series of experiments, but none of them were successful, and the results turned out to be the opposite of what was intended. Then, one day, it dawned on the researchers that weak glue had its own advantages too. The company applied for a patent for the glue, which became an element in the paper stickies that we use today.

The second situation is one where innovation comes naturally. A case in point is Mozart, who began to compose music when he was three. He died at thirty-seven, but left behind a huge corpus of timeless music. The Chinese poet, Li Bai, was inspired to write when he drank, and the more he drank, the better he wrote. People such as Li Bai and Mozart could create effortlessly. As to how they came to possess such ability, it is not for us to know.

The third type of innovation requires agonizing concentration for new ideas to come. The story is well known of how Archimedes tried to determine the amount of gold on a crown. In the end, he figured it out while taking a bath one day. His joyful shout of "Eureka!" left us with the Archimedes' principle of flotation. Many people go through the same process of discovery. After a long period of meditation and experimentation, they finally, as if by chance, get it. This is of course innovation.

The fourth type is most common, and most essential, and does not in any way conflict with the foregoing three. This is captured by the English proverb, "Necessity is the mother of invention." In other words, innovation will occur when there is a necessity for it. I can think of the example of Cao Pi who ordered his brother Cao Zhi to compose a poem in the time it would take to walk seven steps. If he failed to do it, Cao Pi would have him killed. Faced with such necessity, Cao Zhi came up with the famous lines "Being cooked over a fire of beanstalks, the beans weep sadly in the pot . . . "

Talented Innovators Have At Least Five Common Traits

Below, I would like to share my preliminary thoughts on the common traits of those who possess the talent for innovations.

First, I will affirm the importance of a healthy mind and well developed brain cells. This is a necessary but not a sufficient condition.

Secondly, a talented innovator has to have a strong sense of curiosity, and a consuming interest in the work he pursues. This is also a necessary condition.

Third, one has to have a comprehensive knowledge of one's field of work. It will not do for a physicist to have only a smattering of knowledge of physics. This is the reason that I find it necessary for students to engage in the investigation of a broad range of knowledge. Similarly, a poet has to memorize and explicate a large body of poetry before he can write poems of his own.

Moreover, someone with a truly innovative mind has to have an optimistic outlook and a resilient spirit. One cannot just give up at the first setback. Had the stickies researchers that we just talked about stopped after their first failed experiment, they would not have had the accomplishment of today. That is why, without perseverance or determination, innovations would not come easily.

Lastly, whether talented innovators can realize their potential is also dependent to a large extent on the environment in which they find themselves. The environment has to affirm their value, and bestow upon them rewards, be it in the form of name recognition or material gain. I think we can safely assert that in a society that respects innovation, it is easy for people to come up with new ideas and products, and in a society that does not, the opposite is true.

In Paris where I once lived, many streets are short and are named after scientists and artists. Even Gay-Lussac, a scientist not particularly important, has a street named after him. In the minds of the French, this is a sign of respect they pay Gay-Lussac. In Beijing, one cannot find even a single street named after scientists or artists. The example, I hope, will show why innovations have continued to come out of France.

There Are Two Conditions for the Cultivation of Innovative Talent

The next question is whether innovative talent can be cultivated. What are the factors that may facilitate or impede the cultivation of such talent? My answer, of course, is that innovative talent can be cultivated. Otherwise, I would not have agreed to give this speech. However, I do not know what the most important determining factor in the cultivation of innovative talent is. If we knew, our task would be simple. All we would need to do is to cast a model and mass-produce it as in a factory. But innovative minds do not fit into a model, and hence the topic of my speech today: "A closely tended flower will not bloom, but a carelessly planted willow will grow into shade."

I will first take a look at primary education to see how it can be made to facilitate the realization of innovation. Primary school students should attain basic knowledge of science and the accomplishments of human society, and of course, it is necessary for them to learn to read and calculate. We should, moreover, give them time to play or time when they do not have to do anything. Let their imagination roam. Even if they are there to daydream, there is substance in dreams, too, and it would not be a bad thing for them to exercise their imaginations. The first lesson that I learned in primary school was "Come, come, come, let's come to school. Go, go, go, let's go to play." Two things need to be pointed out here: first, you have to learn, and second, you also have to play. Play is indispensable, for it teaches us how to interact with people and energizes us. For some children these days, a more apt description may be "Come, come, come, let's come to school; Go, go, go, let's go to cram school." Their world leaves them no room to think, and the opportunity for innovation is thus reduced.

In secondary school, we expect, on the one hand, the acquisition of more knowledge about human accomplishments in the arts, literature, social thought, science and technology. On the other hand, we should encourage students not to follow a predetermined dogma, otherwise the desire for innovation will be quashed. Students may entertain immature ideas, but as long as they are not completely off the mark, teachers should not in any way undercut them by making fun of them.

In higher education, I think one of the basic premises in cultivating innovative talent is to train students to formulate questions, not just answer them. This is most important.

A person with an innovative mind has to fulfill two requirements in the university and in society. First, he has to place demands on himself and have desires in his mind. Individuality, therefore, is of categorical importance in the process of innovation. Any new idea has to go through the thinking process in an individual's mind. But failure to get along with other people in one's profession or to participate smoothly in team work because of an overblown ego is also an impediment to innovation. Some cultural analysts point to the emphasis on the collective at the expense of the individual in Chinese ethical views; and by contrast, the promotion of individualism and the relegation of the collective in the United States. This, they argue, explains why Americans are more innovative.

I have lived in both societies, and I believe that my views are quite balanced. Collectivism and individualism are present in both societies, and it is difficult to say which of these two peoples are more innovative. Let me give you a simple example: there is no question that Americans tend to lay more emphasis on the individual, but all American children commit to memory a line from John F. Kennedy, the most respected of all American presidents, "Ask not what your country can do for you. Ask what you can do for your country." What kind of ideology is being promoted here? On the other hand, are there not conditions in Chinese society that encourage innovation? Of course, there are. How could the many innovations that I mentioned just now have taken place in China if such was not the case?

The next thing is that an innovative mind has to be knowledgeable. It is perhaps basically accurate to say that teachers are more knowledgeable and correct than students in primary and middle school, but such is not necessarily the case in the university. In China, the idea of respect for teachers and the orthodoxy they represent has grown ever stronger since the Song dynasty. As a result, the presumed superiority of teachers, while not universally true, has come to be taken for granted by teachers and students alike, even in the university.

There is No Set Formula for the Cultivation of Innovative Minds, But There Are Factors That Prevent Them From Being Formed

Is there a set formula for the cultivation of innovative minds? I have already answered that no such formula exists for us to follow. But are there education systems that impede the formation of innovative minds? I believe there are. Allow me to go into this briefly.

I will use the United States as an example. Primary education there is not very successful. Many primary school graduates cannot even read the newspaper, but the schools encourage students to develop themselves, with the result that some of them can even recite passages of Shakespeare. Their mathematical skills are also extraordinary. In East Asian countries such as China, Japan and Korea, you may not find students who are exceptionally bad, but exceptionally strong students are not easy to find either. In the United States, however, while the bad ones are very, very bad, the good ones are truly remarkable. As long as five percent of the students are very good, the results of their efforts are enough to benefit society as a whole.

It is the same way with American high schools. They may

not pay much attention to overall quality, but they provide ample opportunities for the good students to distinguish themselves from the crowd. As for universities in the United States, they can well be the source of most of the innovative minds of the world. In the last one hundred years at least, the United States has given us the greatest amount of innovative talent.

It is my deep conviction that all roads lead to Rome. Thomas Edison did not even finish primary school, but his mother had faith in him, and he was able to continue to learn at home. In the end, he was responsible for many inventions. Similarly, Einstein did not complete high school. Of course, if we followed this method of cultivating innovative minds, the cost would be more than any society, including the wealthiest, could bear. Ours is a knowledge-based society. One cannot hope to become another Edison by not going to the university and staying at home.

There are social factors in China that hinder the growth of innovative minds. One of these is that Chinese tend to praise children for being "*guai* or docile" and "*tinghua* or compliant," which leads them to think that there is no need to do any other thing. As long as they are docile, they will do well. On the contrary, an innovative person, by definition, always impinges upon existing ways and things.

In recent years, the number of university graduates in China has been increasing. It would be a serious misuse of the resources of society if we failed to produce in the university the innovative talent that society needs. The skills and knowledge that students learn at the university nowadays may become outdated even before they retire. The truly useful things to give to students are the desire to learn, the method to learn, a self-motivated and persevering spirit, and a more complete frame of knowledge.

The City University of Hong Kong aims primarily to train professionals such as engineers, lawyers and accountants, but we have a School of Creative Media. This is one of my innovations. We hope to integrate humanistic thought, innovative ideas about the brain, narrative skills and storytelling techniques with the latest digital technologies.

Three Premises
for the Development of Innovation

All societies nowadays are searching for innovative talent. It is the belief of all the latest economics theories that in the beginning and middle stages of economic development, investment should take the lead. The bigger the investment, the more developed the economy. At a higher stage of development, however, what is needed is innovative talent to create new things from existing products, along with services and management systems so that people can work in accordance with these new things.

My personal observation is that East Asian countries have already reached a remarkable level of industrialization. Japan did so a century ago, but it has made little contributions to the important innovations of the last hundred years that we just mentioned, from quantum theory to nuclear magnetic resonance imaging, from venture capital investment to hedge funds. Does the fact that there are more East Asians who readily "toe the line" explain why there seems to be fewer innovations in this part of the world? As people who live under the influence of Confucian ethics, should we not reflect upon the motivation behind our innovations, and find out the factors that limit them?

In my view, for innovation to develop, there has to be first of all a free flow of information and free expression of ideas. Secondly, individual innovations have to be respected. Third, accomplishments in science and the arts and their value have to be determined by experts in the field, or at least by the repeated assessment on the part of many people, and not any individual in high position or by human or divine dogmas.

14
The Brave New World of Learning

Information technology (IT) in education is a topic of fundamental importance to the future of higher education all over the world, and in particular to this region. I believe IT will change the core activities at university – teaching and learning.

Information technology, with the microprocessor as its soul, has become the staple of instructional technology in American higher education. A recent survey, for example, reports that 24 percent of all college classes in America are now taught in classrooms equipped with computers, with some 20 percent of the classes employing e-mail. Peter F. Drucker, the American visionary thinker and management guru, recently told *Forbes* magazine that he believes universities and colleges in the U.S. are beginning to deliver more lectures and classes off campus via satellite or two-way video.

The extent of technology in use in higher education – computer-aided presentations, the Internet, CD-ROMs, video-conferencing, etc. – is not surprising, but the implications will be. What is important about information technology is neither its hardware or software but its disrespect for boundaries and barriers. IT can free us from the constraints if time, distance, and bricks and mortar in learning. Building brickwork and

---

Based on a lecture delivered at Fudan University, Shanghai in April 1997.

stained glass edifices is costly. They don't matter any more in our new learning environment. By using IT, learners can interact with their peers halfway across the world as easily and freely as with those halfway across the classroom or hallway. *Lingua Franca*, a magazine of academic life in the U.S., aptly labels this trend as "classroom in cyberspace" and "education without borders."

IT can also free us from rigid academic boundaries. A major development in higher education in the twentieth century was the creation of infinite new disciplines, each with its own focus, perspective and even its unique methodology of inquiry, that is its own boundary. IT offers the promise of integrating, merging and networking the relationship between academic activities, and removing these boundaries. Some thinkers suggest, as IT proliferates, that there is bound to be a re-alignment of academic disciplines as old ones are eliminated, subsumed or merged with others.

The breaking down of physical and time barriers in learning has heralded the dawn of an era in which learning will move away from the confines of the university campus, and teaching and learning do not have to take place at the same time. Physical distance will no longer be an impeding factor as language or culture is. The Internet, despite its origin in academia, has taken on a life of its own. The Internet, and particularly the World Wide Web, is spawning a myriad of learning opportunities beyond the control of those in the ivory towers. The Web is said to be doubling in size once every six weeks or so and by the end of 1996, fifteen million households in America and twenty-four million worldwide had already had some form of on-line access. In the year 2000, these numbers will be forty million and sixty-seven million respectively. What is even more amazing, is that 59 percent of these households used the Web as an educational resource during 1996 and this shows the Web's potential in an area we have until now largely ignored. This ratio will certainly rise by a few notches in the year 2000.

Now, at the click of a mouse, a learner can read the complete collection of Shakespeare's sonnets, appreciate the fine paintings in the Louvre, tour the Exploratorium in San Francisco, or learn the molecular structure of a new protein being genetically engineered in a biology lab in MIT. The "Brave New World" of learning, however fluid and unstructured it seems to be, is no longer the monopoly of universities as it has been in the past five hundred years or so since the appearance of the *Gutenberg Bible.*

In a knowledge society, as Drucker describes in his book *Post-Capitalist Society,* the controlling resource is neither land nor labor, as it was in the past. It is knowledge. In the new economy, which is increasingly digital in nature and driven by information technology, work and learning are converging. Indeed, the learning part of work looms larger and larger these days. Consider a genetic research scientist working on cloning a sheep or the software programmer developing a search engine for a database system. In each of these tasks, thinking and learning are intertwined in order to achieve satisfactory performance. So in a knowledge-based society, knowledge workers have to replenish their knowledge base every once in a while throughout their life in order to cope with the changing demands of society. We are already living in a knowledge society, more aptly known to us educators as a learning society.

To sum up, IT has made the world of learning truly flexible, interactive and multi-sensory in nature, delivered by the marriage of computers and telephone lines. The promise of IT, however, poses fundamental challenges to the mission of universities. If we wish not only to remain relevant to the changing needs of society but also to shape and lead such changes, I think we have to tackle four pressing issues.

1. Incorporating computer skills as part of the basic literacy requirements' in our curriculum

A study on the use of computers in education in the United States found that two-thirds of the 250 institutions surveyed either have planned, or are planning, to institute computer competency requirements for students. But this number is hardly enough. Yet how many universities in Hong Kong and the Asia-Pacific region have such competency requirements in our curricula?

Not only should universities require basic literacy requirements in verbal and communications skills, and a certain level of professional knowledge in the a chosen area from their students: basic literacy these days should also include computer skills. These embody a considerable understanding of information technology, its dimensions, its characters, its rhythms. Above all, university students should be able to search for, retrieve, organize, synthesize and archive a vast amount of data in different formats (text, graphic, video and audio) for a specific purpose. These skills, I stress, are over and above the equivalent skills expected of the students in the printed world. Central to these skills in the sea of digital information is a clear sense of orientation in cyberspace, a strong ability to track one's own learning footpath, and an unadulterated sense of purpose, that is, the ability to manage one's learning in an entirely different environment. The challenge ahead, for students and educators alike, is to learn how to discriminate and be critical about the vast amount of information that is available.

2. Designing new instructional methods that befit the nature of new technology

Carol Twigg, vice president of EDUCOM, a non-profit consortium in the U.S. dedicated to transforming higher education through information technology, says that "the application of new technology today is simply bolted to old instructional methods." One such example, cited in *Lingua Franca*, is a university in Chicago using "an electronic blackboard" system that can copy what the instructor writes on it directly into the students' laptop computers. All the hardware involved is state of the art but the pedagogical rationale behind it unfortunately is not. It is no more than a souped-up version of "chalk and talk" and hinges on the passive acquisition of knowledge. According to the same magazine, "Souped-up versions of the lecture hall exemplify bolting on."

A better and more innovative approach would be a "studio" approach, where the lecture classes are divided into small groups. Students both work individually with microcomputers and, more often than not, interact and collaborate with one another to acquire a more stimulating learning experience. The advent of IT has called for new pedagogical designs, taking full advantage of the new theoretical developments in cognitive sciences and educational research, that will contribute to an active and constructivist approach to learning. After all, why bother with the new technology if it can't offer something unique, something meaningful to the learners that is way beyond the capability of the existing instructional media. The keywords to watch are: "multimedia", "interactive", "flexible", "instantaneous", and "collaborative", which all adhere to the instructional characteristics of the new technology. But the ultimate litmus test, I think, is the learning outcome. That is, how to use the technology to enhance the learners' experience and improve their learning.

So, instead of worrying if the classrooms and their jobs are vanishing into fiber cables and if the whole campus will be shut down, university educators should start probing for answers to the following questions:

What learning tasks or situations are best suited for the use of technology? How does learning change as new technology permits students to be further removed from the instructor? How can existing pedagogical methods be adapted to the new technology? What new ways can be created to enhance the learning outcome? How can quality in the new, flexible learning technology be kept as high as that in the classroom, which until now has formed the most definitive learning experience for generations of learners? Finally, how should learning outcome be assessed?

A great deal of research has been carried out in these areas, but a clear picture has not yet emerged. Until then, we are still circumnavigating the periphery of a revolution that Drucker describes "as large as when we first got the printed book."

3. Universities should lead in the development of flexible learning technology and opportunities

Universities will slowly lose their monopoly over higher education. As the knowledge economy prospers and requires a lifelong learning commitment from us, the business sector is likely to shoulder a growing responsibility for learning. This view is best espoused in *The Monster Under the Bed*, a stimulating book by Stan Davis and Jim Botkin. The book argues that education, once a purview first of the church, and then government, is increasingly falling into the hands of business because it is business that ends up having to train knowledge workers on the job. Business enterprises will have to become education provider in order to compete in the knowledge-based economy. What universities can do now is to forge partnerships with business, help them plan their employee education, help their employees re-build their knowledge base every once in a while. Universities have the experience, the intellectual assets, and hopefully the foresight and the determination to achieve this.

4. Enthusing students with a willingness to learn

In the "old" economy, the life of an average citizen was divided into a period of learning and a period of earning (your living through work). We went to school and even university to learn a certain competency, and for the rest of our lives our greatest challenge would be simply to keep up with developments in our respective areas. In the new, knowledge-based economy, we, or our students, are expected to re-invent our knowledge base throughout our lives.

E. A. Filence, an American businessman, perhaps sums it up best: "When a man's education is finished, he is finished." So, in the learning society, the first lesson our university students will have to learn is learning is a lifelong pursuit. They will also have to learn how to learn. If we can achieve this, university education will continue its indispensable contribution to society, and universities will survive the information age, with or without a large campus.

15

A Leading University
in the Asia-Pacific Region

Inauguration Speech
at City University of Hong Kong
on May 10, 1996

Chancellor, Chairman of Council, Honored Guests, Students and Colleagues:

Standing here, I feel gratified and excited. Please allow me to be somewhat personal.

When I graduated from the National Taiwan University thirty-four years ago, I was not at all certain that I would later choose teaching as my career. When I first visited Hong Kong thirty-three years ago, I had no idea that I would one day settle in Hong Kong and plan to complete my career here. Yet what has happened, despite my lack of foresight, deeply gratifies me.

In the past thirty-three years, Hong Kong has become a modern metropolis of great importance not only in East Asia but in the world. In the same period, I have followed a career path that has taken me, first as a graduate student and later as a teaching staff, to nine universities in four different countries. It now seems clear that all the wandering and all the work I have done has been to prepare me for the role for which I was just installed to perform. The teacher has heard his calling; the intellectual nomad has found his oasis.

According to the statutes of the City University of Hong Kong, I have the choice to use the title Vice-Chancellor, President, or a combination of the two. In keeping with the prevailing custom of this region and with the consent of the Council, I shall henceforth use the simpler title President.

Having made this choice, I must say that a title is just a title. I have always prided myself in being an intellectual and a teacher. Whatever my title, an intellectual and a teacher I shall remain.

I am excited because of the opportunities and challenges which lie ahead.

To explain, I will speak of the three T's and three A's.

The City University of Hong Kong is simultaneously facing three different transitions, hence the three T's.

First, it is moving, as we all are, from the current millennium to a brand new one. In the first half of the present millennium, individuals and groups that controlled the most land had the most wealth. In the second half of this millennium, following the Renaissance in Europe and the discovery of new routes by the Europeans, what we call Capitalism gradually formed and spread over most of the world. Individuals and organizations that had the most capital wielded the most influence. In the last few decades of this millennium, however, a new and major world transformation has been taking place. In what the American thinker Peter Drucker calls the post-capitalist society, wealth and power no longer come from land, capital, and labor but are based on knowledge. As we approach the threshold of the new millennium, there is increasing evidence that we are moving into such a knowledge society.

As repositories of knowledge and a major source of new knowledge, universities will have a much more important role to play in shaping the future of society than they have had in the past. As a successful modern metropolis, Hong Kong has done well in having established a number of new universities since 1963. As knowledge becomes an ever more important economic resource, Hong Kong's future success as an economic power will depend more heavily on universities. I hope the City University will do its share to usher Hong Kong into the new era.

Second, Hong Kong will face the transition from a British colony to a Special Administrative Region of China. While the transfer of sovereignty will take place overnight, other aspects of this transition will span a much longer period. Hong Kong's old role of being a gateway to China must be complemented by its new role as a front-door of and for China; its people, instead of being outsiders looking in, will have to act on China's behalf, interfacing with the outside. The universities in Hong Kong have a responsibility in preparing its future leaders for the cultural and psychological transition of Hong Kong. Thus, the students of the City University will need to be familiar with the history and civilization of China as well as the West. They must all have a high degree of proficiency in both the Chinese and English languages. The future graduates of the City University should be young men and women who take pride in their own cultural heritage but are well acquainted with other cultures; they should be equally at ease when discussing 红楼梦 (*A Dream of Red Mansions*) and 水浒传 (*Outlaws of the Marsh*) as reading Dickens and Hugo.

The third transition is the transformation of the City University from a former polytechnic into a full-fledged university. It has a diverse student body, a talented and dedicated staff, and an amazing record of rapid expansion coupled with major structural changes. But it has a small and crowded campus. What do I, the new President of this young university, wish to see in the years to come?

I wish to see the City University continue to develop its strengths, acquire a more definite identity for itself, and be recognized as a leading university in this region.

To achieve this goal, let me now speak on the three A's.

The first A is for Academic excellence. Given the rapid changes that have taken place in the University, consolidation and enhancement will be our central tasks in the next few years. Our course offerings and their recently developed infrastructure should be consolidated. The quality of our instructional programs and of our research output must be elevated further. The teaching staff should know not only the subjects they teach, but also how to help the students learn. Taking advantage of the latest information technology, we should adopt more effective teaching and learning methods. Students must be encouraged, indeed, guided to develop the ability and habit of self-learning. Fundamentally, the quality of any instructional program can manifest itself only in the learning experience of the students. We must ensure that it is a rewarding one for the students at the City University.

Research is becoming an important function at the University and I shall do my best to promote research output. Not only do I wish to see more original work published by my colleagues, I would also like to see more significant applied work conducted by our staff.

Whether promoting the quality of teaching or research, the most critical step is to recruit, develop and retain quality staff. It is only through their collective efforts that academic excellence can take hold at the University.

The second A stands for Ambience for intellectual growth.

The university years for a student are not just for the acquisition of a certain set of skills or a certain body of knowledge. They should be a period for developing intellectual curiosity, judgment and leadership skills. I say to our students: "Participate in extra-curricular activities. Learn not only to do things right, but also to do the right things."

To help the students develop such an ability requires a proper ambience. We should adapt our physical plant to generate an environment favorable for intellectual endeavors. We need a campus that promotes scholarship and inspires creativity. We shall persist in asking the Government to allow us to extend our Phase II construction to accommodate our enhanced activities. To foster a genuine campus life, we wish to see the early construction of the badly needed student hostels.

The third A is for Accountability. As a public university, we are certainly accountable to the society that nourishes us. I am pleased to see that our University has already embarked on a series of re-engineering projects. I would like to continue to make our University more cost-effective by deepening the re-engineering process and broadening its scope.

Viewed as a whole, the only relevant measures of our University's output are our graduates and our research results. Everything we do at the university should be in support of teaching and research. Whatever we do that does not contribute directly to these two functions is part of the overhead. The City University will continue to take the lead in being accountable for the resources it receives.

Our University aspires to become a leading university in the Asia-Pacific region. While we are still quite a distance from this goal, I am nevertheless encouraged by what I have seen at the University, and feel certain that we can achieve this vision.

In striving to do so, we can all take inspiration from Yen Yuen 颜渊, the favorite disciple of Confucius. When comparing himself to the ancient sage Shun 舜，he said: "舜何人也？予何人也？有为者亦是。" "What kind of man was Shun? What kind of man am I? If I exert myself I shall become such a man as he was."

The City University of Hong Kong will become a leading university in the Asia-Pacific region if we exert ourselves.

Let's get to work!

16

The World Wide Web

When I began my graduate studies at Stanford University in the United States in 1963, my fellow students and I still relied on the slide rule for computational work. For those of you who have never seen a slide rule, I am holding the slide rule that was my constant companion from 1958 to 1972.

Sometime in 1964, a good friend of mine who was studying electrical engineering at Stanford told me in a small Chinese restaurant that he had just learned how to make an electrical circuit consisting of nearly one hundred transistors on a small silicon chip. He said that this technology was called an integrated circuit or IC. At the time, I was too busy using my slide rule to appreciate the significance of this technology or foresee its impact on our lives.

In 1972, when I was an assistant professor at State University of New York, I spent US$400, or roughly one-third my monthly salary at the time, to buy a newly invented hand-held electronic calculator which contained a microprocessor manufactured with IC technology. This slide rule has never again been used.

Meanwhile, my friend at Stanford has in the last thirty years worked in what has become known as the Silicon Valley, the Mecca of the high-tech industry in the world. He is now Senior Corporate Vice-President at the Intel Corporation and has been responsible for the development and production of the 286, 386, 486, Pentium and Pentium II microprocessors. The Pentium II processor that came out in 1997 has eight million transistors on a single silicon chip with an area of only about one half of a square centimeter. Back in 1964 it was the possibility of putting some one hundred transistors on a single chip that excited him!

A few months ago, this friend was in Hong Kong and had dinner with me. This time he told me another piece of news. He thinks that in ten to twelve years the new generation of processors will have one billion transistors on them, a hundred and twenty-five times more powerful than the most powerful Pentium II processor we have now. Imagine what a PC equipped with such a powerful processor will be able to do!

Ever since the first electronic computer was made in 1946, the world has been going through a digital revolution. The digital revolution is not merely the result of the amazing progress in microprocessors and computer architecture. Software development and networking technology also have played an important part in this revolution. The Internet, when it was started in 1971, connected only twenty-three computers in the United States. Today, the Internet serves over twenty million computers spread over almost every corner of the earth.

In 1994, when I was Dean of the School of Engineering at the University of Pittsburgh, a staff member asked me whether our school should mount a homepage on a new Internet feature called the World Wide Web. I had to ask him what a homepage was. Now, only four years after the appearance of the World Wide Web, terms like "homepage" and "website" have become household words all over the world. There are now over five million websites on the Internet, many with both images and sound in multimedia form. The volume of traffic on the Internet roughly doubles every three months. When such traffic includes data that combine texts with three-dimensional graphics and intricate sound accessible on a home computer, the new modalities for learning, entertainment, personal and business communication are mind-boggling.

In the short span of one generation, the enormous power of modern science and technology has revolutionized the way we live and work. In a broader context, however, the digital revolution that has sprung from progress in microelectronics, telecommunication, computer networking and software technologies is but one aspect of the new world we have created. Other advanced technologies, such as biotechnology, new materials, automation and robotics, and space and aviation technologies will surely have their impact on our lives in the next few decades. To borrow the title of Aldous Huxley's famous book, we will live in a "Brave New World" of high technology.

From a wealth creation point of view, it has been estimated that eighty per cent of the economic growth in the U.S. in the last two decades has come from innovation in high-tech industries. Although no one can predict with any certainty where new wealth will come from in the next twenty years, it is certain that the societies that can take full advantage of science and technology and can generate new products and services accordingly will have a competitive edge over those that cannot. For the long-term well-being of Hong Kong, it is essential that we educate our youth to master modern science and technology and at the same time develop their creative abilities. Our Chief Executive has emphasized this point in both of his policy addresses; it is encouraging that the Chief Executive's Commission on Innovation and

Technology envisions Hong Kong to be an innovation-led, technology-intensive economy in the twenty-first century.

The American thinker Peter Drucker says "The best way to predict the future is to invent it." Hong Kong must invent its own economic future by investing in education, giving due emphasis to science and technology, encouraging innovation, working in even closer collaboration with the mainland and developing further our international links.

We at City University are well positioned to make significant contributions to these endeavors. We have in the past few years established a very strong record in research and innovation, thanks to the first-rate researchers and state-of-the-art facilities at City University. Our staff have also been working closely and fruitfully with industry and commerce. More fundamentally, we shall do our share to develop the human talent pool that will be needed to make Hong Kong more competitive in the future. In this respect, I am pleased to say that our new School of Creative Media, which opened its doors only in September, is already very popular with the students. In a few years, we shall graduate young talents who not only can master the ever-evolving multimedia technology but also combine the best available technology with creativity. We envisage our graduates as founders of many yet unborn enterprises in multimedia production, whether for education, entertainment, advertising or general communication. The potential for the multimedia industry is enormous and City University hopes to help develop this potential, thereby creating new wealth for Hong Kong and China.

17

Extensive Study, Accurate Inquiry, Careful Reflection and Clear Discrimination

Excerpts of the 2000 Congregation Speech

About ten thousand years ago, agricultural life in terms of planting and animal domestication began to spring up in some parts of the world. This was the beginning of human civilization.

About a thousand years ago, several civilizations had reached a fairly high level of sophistication. Yet no one in Europe had any knowledge of the Americas or Australia. In fact, people generally believed that the earth was flat until about five hundred years ago.

About a hundred years ago, the steamship, telegraph and railway had become commonplace and most educated persons would have some knowledge about the various continents and countries in the world. Shortly after that, the telephone, automobile, motion picture and airplane became popular in many parts of the world. About fifty years ago, television and computers began to play an important part in modern life.

When our University was founded sixteen years ago, the digital mobile phone and the World Wide Web did not yet exist; nor did many people have access to fax machines and personal computers. Today, almost all of us here have a mobile phone with which we can reach another person in the far corner of the world instantly. Many people in Hong Kong and most of the world watched on the Internet the minute-by-minute returns of the American presidential election earlier this month. In this sense, the world has become one and the Latin motto that is on every American coin, *E Pluribus Unum*, namely, *From many to one*, is acquiring a new meaning.

What does this mean to you, the graduates, and to all of us?

First, it means you must be able to adapt to rapid changes in order to cope with modern life. The only way to be able to do so, paradoxically, is to have a firm hold on the fundamentals which do not change much. The adage from 《易经》 (*The Book of Change*): "天行健，君子以自强不息" ("The celestial bodies are regular in their motion, so should the superior man labor unceasingly to strengthen his own character") has benefited millions of Chinese scholars in the past two thousand and five hundred years and it is still very applicable and useful today. The universe is in perpetual motion and things never cease to change. Graduates, to live and succeed in an ever changing world, you must seek to have a firm knowledge of the fundamental nature of the world and never cease to improve yourself.

Second, in this complex world with so much information, it is essential that we learn to discard irrelevant information and acquire genuine knowledge that can help us make sound judgments. There is so much news in and about the world now, but any person can easily be dazzled by this abundance of news rather than be enriched by it. Many people throughout the world have learned the details of the vote counts in the recent American presidential election. How many of us have gained new insight into the democratic process? We all know much about the lives of the British royal family who live in a very different way from ordinary citizens. But how many of us know the way people live and the nature of the conflicts in the south of the Philippines and in East Timor, both of which are in our own region? We are all concerned about environmental hazards to health. But can we rationally compare the magnitude of the hazards of ozone depletion in the upper atmosphere, exposure to electro-magnetic waves from using mobile phones and cigarette smoking? Last month, a patient in the U.K. received an implanted constant-flow pump to replace the action of his failing heart, which normally generates pulsating blood flow in the body. He now lives and works but has no pulse. What does this say about vital signs of life and about heart-throbbing experiences?

Once again, I would advise you to learn more about the fundamentals of physics, chemistry, biology, geography, history and, of course, the most fundamental of them all, languages and mathematics. Only when you have a good knowledge base can you absorb new knowledge and learn how to make a proper judgment in a new situation. In this regard, we can learn from the teaching in 《中庸》 (*The Doctrine of the Mean*): 博 学 、审 问 、慎 思 、明 辨 (Make extensive study of what is good, accurate inquiry about it, careful reflection upon it, and clear discrimination of it).

18

After Extreme Bad Luck Comes Good Luck

Excerpts of the 2001 Congregation Speech

City University takes pride in being "in the City, of the City and for the City." As Hong Kong is facing tremendous challenges as well as excellent opportunities, we should strive to do our share in helping Hong Kong to shape its future. However, Hong Kong does not exist alone in isolation from the Chinese mainland or the world. In this sense, I am reminded of a famous couplet by a scholar in Ming Dynasty: "风声、雨声、读书声，声声入耳；家事、国事、天下事，事事关心。" (The sound of wind, the sound of rain, and the sound of reading – every sound comes to my ears; the affairs of my family, the affairs of my country, and affairs of the world – all affairs are concerns in my heart). Thus in my talk, I will first discuss the broad context in which Hong Kong and City University exist and then report briefly on what City University has done to serve Hong Kong. Finally, I will offer some advice to the graduates.

At the dawn of a new millennium, the world is in an age of globalization, with increasingly freer flow of information, capital, goods and even jobs. It is also a world of strong contrasts between the rich developed countries and the poor developing countries. Strife and conflicts threaten peaceful development and progress for humankind. The recent terrorist attacks in the United States and the subsequent war in

Afghanistan, protracted conflicts between India and Pakistan and the Israelis and Palestinians, tensions around Iraq, in Indonesia and the Philippines all give proof that humanity has not advanced to a stage where we can live in harmony with one another.

The world now has about six billion people, divided into some two hundred countries and more than six thousand ethnic-linguistic groups. Each of these groups has its own history, language and culture. Just as bio-diversity is a good thing in ecology, so should cultural diversity be regarded as a wealth for the human race. So we should learn to understand each other and respect the differences. Yet, even with all the newspapers, radio and television networks, mobile telephones and the Internet, misunderstanding or lack of understanding still exists today in the most developed societies about some of the most numerous and important groups in the world. For example, there are more than one billion Muslims in the world and they form the dominant social group in more than fifty countries or regions and are present in more than one hundred countries or regions, including the Hong Kong SAR. But how many non-Muslims have even a rudimentary knowledge of Islam or a fair understanding of this important religious and social group? Likewise, both China and India have over one billion people. Regrettably too few non-Chinese and non-Indian people know much about the histories, languages and cultures of these two nations with such ancient and brilliant civilizations.

Let me say a few words about China. With the arrival of the Italian Jesuit Matteo Ricci in China some four hundred years ago, China started the process of first learning about and then learning from Western countries. In the nineteenth and twentieth centuries, China was repeatedly humiliated and deeply traumatized. It is now finally going through a modernization process that has a strong probability of success. But at the same time, China faces tremendous challenges. First, it must feed and educate a population of 1.3 billion people on a very scarce resource base in a large and unevenly developed country. Second, it requires a political-economic system that has the authority as well as efficiency to allocate resources according to the needs of the various regions to ensure stability and balanced development. Third, in order to be competitive in the global knowledge-based economy, it must go through industrialization and the digital revolution simultaneously. Fourth, it must minimize the detrimental impact of economic development on the country's environment. Fifth, it must find a proper balance between almost complete urbanization, as in the United States, and its present population distribution with more than 70 percent of the people living in rural areas. How China meets these challenges will be key to its achieving sustained development and enhanced individual liberty, rule of law, social justice and democracy. At this important juncture, it is fortunate for the Chinese nation that the current Government insists upon an open and outward-looking development strategy and has decided to link China's modernization to the rest of the world by joining major international organizations such as the World Trade Organization.

Let us now look at Hong Kong in this context. We in Hong Kong straddle the developed world and the developing world, and stand in the cross currents of the cultures of the East and the West. Our physical infrastructure, legal system and other social institutions put us in a unique position to contribute to China's modernization by serving as a conduit of goods, services and ideas while further developing the Hong Kong SAR economically, socially and culturally. Hong Kong should continue to be a window for China and a bridge between the Chinese mainland and other countries and regions.

With this backdrop, let me now return to City University and how it serves Hong Kong's needs.

At the Congregation held in December 1996, I said, "While we have every reason to be optimistic about the future, we must not overlook the challenges ahead. Among the many challenges one can think of, the one with the greatest long-term effect is the education of our young. As Hong Kong gradually integrates with the expanding economy of the Mainland and at the same time participates more fully in the global economy, the demands on the knowledge and skills of our workforce will indeed be very great." While I could not foresee in 1996 the Asian financial crisis a year later and I certainly did not have any clue of a world-wide economic slowdown to be exacerbated by terrorist attacks in the United States five years later, my thoughts about educating the young and having a workforce that would face the challenges of a knowledge-based economy are certainly pertinent today.

Predicting what City University would be like in five years, I said, "It will become the hub of quality university education in Hong Kong and will produce a large number of broad-minded, well trained and dedicated professionals. The new professionals will have gone through our new credit-unit system that will allow them to have a broader knowledge base than their predecessors . . . and they will benefit from the tutelage of an increasing number of renowned scholars at the University." To judge the truth of this prediction, allow me to provide some facts and figures. While the number of academic staff has held steady at around nine hundred, we now have sixty Chair Professors, twice as many as in 1996. We have significantly strengthened our academic programs, at both the undergraduate and postgraduate levels, in a number of disciplines, including Electronic Engineering, Environmental Biology, Language Science, Materials Science and Mathematics. The evaluations conducted by the University Grants Committee have attested to the excellence in these areas. We have also excelled in applied research and won a large share of grants from the Government's Innovation and Technology Fund. In addition, City University has encouraged staff to develop their intellectual property rights and set up their own companies; one such company is now publicly traded in Hong Kong.

In 1998, we started the School of Creative Media, where students learn to create and produce content in the digital multimedia form. I am certain that the graduates of this School will in time help to build a robust multimedia industry in Hong Kong.

To help strengthen Hong Kong's role as a meeting point for Chinese culture and other cultures, we have established the Center for Southeast Asian Studies and the Center for Cross-cultural Studies. The importance of these centers has been amply underscored by the recent developments in international affairs.

Of course, all of today's graduates are familiar with the English Language Center, the Chinese Civilization Center and the courses they offer. The beneficial effects of these courses will become increasingly clear to you as time progresses and I am sure you will appreciate them even more as you advance in your careers.

All of you should also be familiar with the various programs we have initiated to help our students develop their leadership potential, work ethic, and other attributes. In this regard, I am sure you are as proud as I am that City University's sports teams have twice won the overall championships in both men's and women's events.

Dear Graduates, I would like to offer some of my thoughts to you specifically.

You will soon step up and down this stage. It is an act that will forever link your name with City University of Hong Kong. Yet, the education City University has given you is only a key to knowledge and to your professional life. You must open the door to the palace of knowledge and find the secret to success yourself. Today you are proud to be graduates of City University. I hope tomorrow City University will be proud of you as our alumni.

The future holds enormous promise despite global political and economic difficulties at the present time. To paraphrase Dickens, you are starting your professional life in the best of times and in the worst of times. Yet, whatever the times, you must learn to anticipate and overcome adversity and be resilient. In this, the ancient philosophers of China can be a great source of inspiration, and I urge you to reflect on the meaning of "寒来则暑往，暑往则寒来" (The cold goes and the heat comes, the heat goes and the cold comes); "乐极生悲" (Too much joy leads to sorrow) and "否极泰来" (After extreme bad luck, comes good luck).

19

Embrace Changes and Exert Yourselves

Excerpts of the 2003 Congregation Speech

Forty years ago, I left Taiwan for graduate studies in the United States. At that time, my parents were working in Ethiopia for the World Health Organization. So I left Taipei and visited my parents in Ethiopia, and then flew to the U.S. East Coast by way of Europe, and finally reached my university on the West Coast by taking a bus all the way across the continental United States. We may recall that Columbus, in his effort to find the best route from Europe to China, landed on American shores in the end. My best route from Taiwan to California, on the other hand, took me across three continents around much of the globe – Asia, Africa and Europe.

On that journey forty years ago, Hong Kong was my first stop. A friend from my secondary school days, who lived in Hong Kong, met me at the airport to take me to his home. My friend's mother gave me a warm welcome, saying: "Welcome to our house. There is, however, one hitch: you can only use one basin of water for each day." At that time, Hong Kong's infrastructures were not very good, and there was a serious problem in water supply. In the summer of 1963, tap water supply was restricted to a four-hour period, once every three days.

From Hong Kong, I continued my journey, stopping over in Bangkok and Bombay before I landed in Beirut, the capital of Lebanon. Lebanon was once a French colony, and Beirut was often known as "Little Paris." There, I learned of the sea voyages and history of the Phoenicians, who invented the phonetic alphabet, and I toured and admired the site of a Roman temple at Baalbek. At that time, the U.S. marines had just withdrawn from Lebanon, and I had my first exposure to the political tensions of the Middle East.

From Lebanon, I flew by way of Cairo and Addis Ababa, the capital of Ethiopia, to Gondar, an old city in the country's north where my parents worked. Ethiopia was a kingdom in East Africa with a three-thousand-year history, where conversion to Christianity preceded that of Rome. I had a chance to visit a Christian church with a history of more than a thousand years. I also visited Falasha Jews, who were said to have inhabited in Ethiopia for all three thousand years and whose appearance and way of life were no different from the other Ethopians.

In August, having left Addis Ababa, and after making stopovers at Khartoum in Sudan, Athens, Rome and Zurich, I arrived in Paris, a great city I had admired since my early years. I finally saw the Mona Lisa in the Louvre, whose mysterious smile had mesmerized countless viewers. I also went to La Tour Eiffel and the Arc de Triomphe. During my three days in Paris, I spoke a lot of French, but I discovered that among the people around me, many were English-speaking American tourists.

From Paris, I flew to New York and checked in a room in the Manhattan YMCA for US$3.50 a day. I took the opportunity to climb up the Empire State Building, visited the Metropolitan Museum of Art, and took a stroll along the Fifth Avenue and Broadway. My initial impression of New York was that it was a city of immigrants and tourists, and you hardly heard people speaking pure American English!

Then I caught a long-distance coach in New York and, after a 72-hour-long grueling journey, I reached Stanford University in the Bay Area of San Francisco at last.

Soon, in the library at Stanford, I had the opportunity to read a variety of newspapers and magazines, including those from Mainland China, and I began to have some understanding of world affairs and the racial relations in America. At the time, the Reverend Dr Martin Luther King, Jr. had just delivered his famous "I Have a Dream" speech on the steps of the Lincoln Memorial in Washington, D.C. Many white Americans were moved by him, so they spoke out against racial segregation and supported the Civil Rights Movement. But there were also many others who could not accept his point of view. As a result, race relations in the country at that time were very tense.

One day, around noontime, in the late fall of 1963, I noticed a sad mood gripping the campus, some students even sobbed. It turned out that President John F. Kennedy had been assassinated that morning. That terrible event shook the world, and the facts surrounding the assassination are still murky today.

How time flies and history ebbs and flows! It seems that forty years have passed in the blink of an eye.

Those Asian and African nations that had just shed the shackles of European colonialism in those days have not yet found solutions to many of the problems left by their colonial past forty years after. Today, ethnic and tribal wars are frequent; poverty and corruption still abound everywhere we turn our eyes.

During the same forty years, the United States has welcomed into her arms millions of immigrants from all over the world, and has improved the country's race relations. In the forty years, there are three Secretaries of State who are naturalized citizens born in foreign countries. Today, the U.S. Secretary of State and the President's National Security Advisor are both African-Americans. What an open and liberal-spirited nation we see! And yet, we may ponder why has a country that stresses freedom and human rights within its borders resorted to military force so many times around the world in the past forty years? And how can it reconcile, as its national policies, pluralistic and all-inclusive ideals at home with unilateralism abroad?

As for China, it has learned hard lessons from various political campaigns. In the past twenty years, China has pursued reform and openness, focusing its energy and attention on economic development, adopting science, technology and education as its national policies in the hope of bringing revitalization of the Chinese people. Externally, China advocates peace and development; it is playing an increasingly important role in world affairs. Compared with forty years ago, we should be confident and feel proud of being Chinese today.

Let us now turn and take a look at Hong Kong in the year 2003. Hong Kong is by any standard one of the world's most modern, most affluent and most prosperous metropolitan cities. With their hard work, the two generations before us have pushed those days of no subway system, no cross-harbor tunnels and no adequate water supply out of our collective memory and turned them into traces of a bygone past. Now we are part of a prospering China, and we are at the main gateway of South China to the world. We can contribute to the modernization of China, and at the same time realize our own growth in this exciting historic development. To create a better future, we must resume the striving spirit of our past and use the principle of "One Country; Two Systems" to our advantage, dedicating ourselves to the future of Hong Kong while always bearing in mind the interest of the country.

Yang Liwei, the astronaut in the Shenzhou 5 spaceship, has brought honor to the country and brilliant accomplishment and a sense of personal satisfaction through many years of hard work and practice. In the past forty years, American and Russian astronauts have also toiled long and hard, and made enormous sacrifices for their countries in fulfilling their duties, but they have attained personal success and gratification, too. We can draw some inspiration from these examples: without working for the interest of the society at large, there can be no personal success.

Some have said that Western nations put priority on the individual, while in East Asian societies collective interest always takes priority over the individual. Let us set aside the question of the validity of such simplistic generalizations. Even if such a statement had something valid to it, Hong Kong as the meeting place of Chinese and Western cultures should fuse the two sets of values and find the best reconciliation of collectivism and individualism.

In fact, although Chinese culture is heavily influenced by Confucian thought and puts relative emphasis on a sense of the collective, the mental state portrayed by many poets and painters by drawing on Zen Buddhism and Taoist philosophy are also quintessentially Chinese. A famous poem by Chen Zi'ang, an early Tang poet:

> *Where, before me, are the ages that have gone?*
> *And where, behind me, are the coming generations?*
> *I think of heaven and earth, without limit, without end,*
> *And I am all alone and my tears fall down.*
> *How can this be taken as a display of collective spirit?*

Although the basic values of modern European and

American societies are inclined towards individualism and personal interest, surely there have been uncountable examples of altruism and unselfish sacrifices, like the American and Russian astronauts mentioned above. The motto of the U.S. Military Academy at West Point is: "Duty, Honor, Country." How can one say that this articulates the yearning of individualism?

I have lived and worked both in East Asia and in the West over the past forty years. What I come to realize deeply is this: no matter where you live, it is only by helping one another in a community with a strong sense of solidarity can a society make progress and allow the individual to grow to the full potential. Only by reconciling and unifying individual interests and the interests of the community can a society become more lively and prosperous, and each individual in the society can feel relaxed and do his best.

Dear graduates, as you are about to commence your career or to open another chapter of your life, I have high expectations of you. From where you are standing today, facing challenges in Hong Kong, China and the world, you must work hard, and you will have your achievements.

I would like to offer you this quote, which I learned in 1963, from the late U.S. President John F. Kennedy: "Ask not what your country can do for you; ask what you can do for your country." The bright future for Hong Kong and China lies in every effort you make in the interest of Hong Kong and the motherland. And in the bright future of Hong Kong and China you will find rich rewards both in your spiritual well-being and in your career.

20

Picture Out of Your Life

Excerpts of the 2005 Congregation Speech

Many of you have probably read or heard of *A Tale of Two Cities*, a well-known novel by the English author Charles Dickens. If I had to choose a title for my speech today, I would call it *A Tale of Two Murals*.

Let me begin by telling you where the two murals are. If you entered the University today through the tunnel connecting Festival Walk and CityU, chances are you also came through the Academic Building by the doors directly facing the tunnel. The two murals, which measure ten meters by four meters each, are on the two walls next to this entrance. They are the work of a renowned artist from Beijing (Professor Yuan Yunsheng) who has lived and worked for many years in the U.S. In fact, both Harvard University and Tufts University in Boston commissioned this artist to create mural paintings for their campuses.

The reason I chose to talk about these two murals today is I think this work and the process of its creation symbolize the spirit of our University. Allow me therefore to say a few words on this matter.

In 1994, on the eve of being officially named City University of Hong Kong, the University community engaged in a consultation process to determine the Vision Statement. What was eventually approved by the Council states: "City

University of Hong Kong aspires to be internationally recognized as a leading university in the Asia-Pacific region." Despite the rapid changes that have occurred in Hong Kong, China and worldwide, this ambitious yet realistic statement remains a constant source of inspiration.

In the Strategic Plan for 1997–2002, which was endorsed by the Council in 1997, we defined our "ideal graduates" as qualified, competent professionals who must be "proficient communicators equipped with a range of disciplines and skills, computer literacy, as well as language proficiency" and who have the "ability to think quantitatively and analyze problems critically". To achieve such a goal, we said it was the University's responsibility "to expose our students to the wider world of scholarship in the arts and sciences, and to their culture and history."

In the Strategic Plan for 2003–2008, we summarized the scope of our work by declaring that "the University is positioned along the axis that links professional education with applied research." We said we were making CityU's education special in two respects. "Firstly, we are conscious that our graduates must find their way among the cross-currents of diverse cultures. To prepare them for this challenge we emphasize the value of studying Chinese culture and social life; the dynamics of regional developments, especially in the Pearl River Delta; English language acquisition; and exposure to other cultures. Secondly, we focus on the dynamic linkage between the new technologies revolutionizing our world and the creativity and imagination required to put them to exciting and profitable use."

Consistent with the ideals and educational philosophy espoused in the two Strategic Plans and taking full account of the Role Statement given to our University by the University Grants Committee, the Council adopted in August this year a

Mission Statement which enunciated our tasks with this preamble: "The mission of City University of Hong Kong is to nurture and develop the talents of students and to create applicable knowledge in order to support social and economic advancement".

Ladies and Gentlemen, so far I have given you an account of how the goals, major tasks and the underlying philosophy of our University have evolved and how they have been articulated in official documents. You would be justified to ask, "What about the two murals?"

Let me briefly describe them for you.

The first painting depicts a Chinese scholar from ancient times caressing a *guqin*, the most antique and unique Chinese musical instrument. Shown in a somewhat contemporary Western motif but with an unmistakable Chinese aura, the scholar is in a deeply contemplative mood. To me, this painting represents the human spirit in intellectual pursuit. The second painting, also a fusion of contemporary Western and traditional Chinese styles, portrays a man trying to soar into the sky. The man's face and his eyes radiate excitement and exaltation, but he is a far cry from the astronauts we see on our television screens today. Any person from our own era knows that this man is never going to fly high or far. Yet this picture is not a creation of the artist's imagination, but his re-creation of a real story that happened during the Ming Dynasty. A man by the name of Wan Hu tried to fire himself into space using simple rockets first invented during the Sung Dynasty. Again, this painting pays homage to the noble human spirit and extols the value of technological innovation.

Both murals exhibit a graceful combination of traditional Chinese cultural elements with Western means of expression. They are a genuine fusion of East and West and a wonderful example of the synergy between the old and the new. Their presence at the entrance to CityU's Academic Building demonstrates our commitment to modern professional education, buttressed by deep-rooted humanistic values. The paintings graphically illustrate the technological competence and cultural literacy that underpin our persistent drive to "nurture and develop the talents of students" and to "create applicable knowledge."

In this tale of two murals there is a side-story which I think is also quite revealing. The committee responsible for procuring paintings for the two walls decided to commission the author of these paintings after an examination of his achievements and a face-to-face interview. The committee members explained what CityU stood for and expressed a wish for him to incorporate these ideas into his work, but, at the same time, gave him complete freedom to paint the murals.

During the eighteen months it took for the paintings to be completed, an unfortunate accident occurred. One night when the artist was working alone in his studio, he fell off a ladder, breaking his ribs and severely damaging his right arm. He was hospitalized and had to go through rehabilitation. At the age of sixty-five he was frustrated but undaunted. He told me later that he treasured the opportunity to work for a client who granted him total artistic freedom so much that he redoubled his efforts once he was able to climb the ladder again. He said he also gained more insight into Chinese philosophy and culture during this unexpected interlude. In the end I think he probably created a more profound piece of art because of his mishap.

This story confirms two basic beliefs held dear at CityU. First, no great intellectual work can be accomplished without the dedication of its author. Second, intellectual creation is most effective when there is intellectual freedom. As an educator, I have subscribed to these notions all my professional life and am extremely proud that CityU has benefited from these basic values.

Since we have been talking about technological competence, cultural literacy and creativity, I would like to offer you my own reflections on these subjects. Creativity does not come out of thin air; it usually springs from a foundation of relevant knowledge and a given cultural environment. People who do not have a strong technical and cultural background are unlikely to create new knowledge or find new solutions to real problems. Conversely, there are countless examples of people who are creative because they have the self-confidence derived from their cultural awareness and can see connections in the seemingly unconnected due to their broad knowledge base.

Exactly one hundred years ago, Albert Einstein published his theory of relativity. We all know he was a very creative physicist with a broad range of interests and his work helped reshape the world. Yet he was also a well-read man and a competent violinist. By his own accounts, he owed his success in physics to his diverse intellectual stimulations, which included music and literature.

In the same year Einstein published his theory of relativity, China abolished the 1,300-year-old *Keju* (科舉) examination system and launched a modern education program in which mathematics, science, geography as well as Chinese classics were taught. Einstein's work was probably unknown to the Chinese education reformers at the time, but their decision to favor a broader curriculum proved farsighted and had a far-reaching impact. It was consistent with Einstein's key to success. That China has taken huge strides in the twentieth century and is narrowing the gap between it and the more developed nations in the world can probably be traced back to decisions taken in 1905.

Dear Graduates, the fact that I have in one speech talked about CityU's educational philosophy, the story of the two murals, the success of Einstein, and the introduction of a new education system in China may seem unconnected. But are they really? I will leave the answer to you.

What I can say now is that having a broad background is like having many colors for the painting of life.

Dear Graduates, in front of each of you is a clean canvass. CityU has prepared you with the colors. Use them creatively. I am sure you can paint a beautiful and brilliant picture out of your life.

21

The World Is Flat

Excerpts of the 2006 Congregation Speech

City University's pride is derived from its ability to transform young students into successful professionals. As you start your careers, your chance of becoming a successful professional and producing creative work depends how you adapt to the deepening and accelerating globalization process.

That the earth is like a globe was known to Christopher Columbus. Historically speaking, globalization started with his arrival in the Americas in 1492 and Vasco da Gama's rounding the southern tip of Africa to reach India in 1498. But the globalization process, as we know it now, has become more profound and more rapid in the last twenty years compared to the previous five hundred years.

I lived in North America for many years and have been to India on several occasions, including spending one month there this year. What I can say is that North America and India are now intricately linked by information technologies and are interdependent. When an American consumer makes a phone call to place an order, make an inquiry on a bank account or plan a trip, the person who provides the answers at the other end of the phone is likely to be someone sitting in an office in India. In 2004, more than one million Indians were employed by multinational companies to do office work, ranging from answering calls and doing accounting to conducting laboratory tests and developing software. More and more Indian engineers and radiologists are doing engineering design and making diagnostic reports on X-ray films or CT-scans.

Being in Hong Kong, we all know from the volume of trade that the economies of China and the U.S. are closely intertwined and mutually dependent.

Two years ago, my wife and I were eating in a good French restaurant in a historical city in France. The owner and chef told us that he could no longer compete with McDonald's, Pizza Hut and the like. So he had just signed a contract to work for a hotel in Hangzhou, China. This is yet another example of globalization.

But moving any service that does not require personal contact to a place of lower cost will not be limited to the types of work I cite here, nor will India and the Chinese mainland be the only places where such work is destined.

By any standard, Hong Kong is a high-cost city. So it is very likely that some of the professional work that is now performed in Hong Kong will move to India, mainland China, Malaysia, the Philippines and other places.

Last year, the Pulitzer Prize-winning American writer Thomas Friedman wrote a book with a title that would surely startle Columbus: *The World Is Flat*. Friedman argues that the twenty-first century will be remembered for a new age of globalization – a "flattening" of the world. The explosion in technologies and resources has connected all over the planet, leveling the playing field as never before, so that each professional worker is potentially an equal competitor. Who wins and who loses depends on the ratio of the knowledge and creativity a professional can offer to the salary he or she demands.

Quite apart from pay, the tools used in professional work

continue to change. When I was an engineering student in the early 1960s, I had a slide-rule with no mechanical or electronic parts to make all my engineering calculations. When I taught engineering in the early 1970s, my students began to use hand-held electronic calculators to do the computations. Today, no self-respecting engineer uses such simple instruments for professional work; they are only used by sales people in tourist shops to bargain with customers.

And it is not only the tools and materials that have changed. Many professions have fused together; others have disappeared. For many professions, the basic methodology has also changed. Some principles once held sacred are being re-examined. The only thing that has not changed and will not change is the human quest for new knowledge and the human spirit for improvement.

Dear graduates, this is the field of competition you are entering. It is as if you were selected to compete in the 2028 Olympic Games, but you do not know in what event you will be asked to compete, nor will you know the rules of the event until 2027. But do not let that prospect scare you. You have been given the necessary preparation at City University for this type of competition. Concentrate on what you can do to add to your overall strength and develop as broad a background as you can. Adapt to new conditions willingly and swiftly. Most importantly, rely on your cultural heritage. Remember the two Chinese proverbs that most of you learned as children: "学而后知不足" (Only after you learn will you realize there is much more to learn.) and "自强不息" (Never cease to strengthen yourself.)

Dear Graduates, you will be in your prime in 2028 and probably be on an ascending curve in your career development. By then City University, your alma mater, will have become a leading university in this region of the world, with hundreds of thousands of alumni contributing to Hong Kong, mainland China and other regions of the world.

The HKSAR will have been in existence for more than thirty years, and, I dare say, will have integrated itself well in the economic, social and cultural fabric of China.

I know our University, Hong Kong and you will all do well in 2028. Dear Graduates, as President of City University of Hong Kong, I give you my best wishes and hope to see all of you win in your own Olympic Games in 2028.